MW01488697

Contaminant Transport Modelling in Heterogeneous Porous Media

Sina Borzooei

Contaminant Transport Modelling in Heterogeneous Porous Media

Comparison of analytical, numerical and stochastic methods in one-dimensional contaminant transport modelling

LAP LAMBERT Academic Publishing

Impressum / Imprint

Bibliografische Information der Deutschen Nationalbibliothek: Die Deutsche Nationalbibliothek verzeichnet diese Publikation in der Deutschen Nationalbibliografie; detaillierte bibliografische Daten sind im Internet über http://dnb.d-nb.de abrufbar.

Alle in diesem Buch genannten Marken und Produktnamen unterliegen warenzeichen-, marken- oder patentrechtlichem Schutz bzw. sind Warenzeichen oder eingetragene Warenzeichen der jeweiligen Inhaber. Die Wiedergabe von Marken, Produktnamen, Gebrauchsnamen, Handelsnamen, Warenbezeichnungen u.s.w. in diesem Werk berechtigt auch ohne besondere Kennzeichnung nicht zu der Annahme, dass solche Namen im Sinne der Warenzeichen- und Markenschutzgesetzgebung als frei zu betrachten wären und daher von jedermann benutzt werden dürften.

Bibliographic information published by the Deutsche Nationalbibliothek: The Deutsche Nationalbibliothek lists this publication in the Deutsche Nationalbibliografie; detailed bibliographic data are available in the Internet at http://dnb.d-nb.de.

Any brand names and product names mentioned in this book are subject to trademark, brand or patent protection and are trademarks or registered trademarks of their respective holders. The use of brand names, product names, common names, trade names, product descriptions etc. even without a particular marking in this works is in no way to be construed to mean that such names may be regarded as unrestricted in respect of trademark and brand protection legislation and could thus be used by anyone.

Coverbild / Cover image: www.ingimage.com

Verlag / Publisher:
LAP LAMBERT Academic Publishing
ist ein Imprint der / is a trademark of
OmniScriptum GmbH & Co. KG
Heinrich-Böcking-Str. 6-8, 66121 Saarbrücken, Deutschland / Germany
Email: info@lap-publishing.com

Herstellung: siehe letzte Seite /
Printed at: see last page
ISBN: 978-3-659-50378-8

Copyright © 2014 OmniScriptum GmbH & Co. KG
Alle Rechte vorbehalten. / All rights reserved. Saarbrücken 2014

CONTAMINANT TRANSPORT MODELLING IN HETEROGENEOUS POROUS MEDIA

By

SINA BORZOOEI

Department of Civil and Environmental Engineering
J.N.T.U.H COLLEGE OF ENGINEERING
Division of Geo-Environmental Engineering

1

ACKNOWLEDGEMENTS

First of all I would like to express my gratitude to all who have directly and indirectly assisted in this work. First of all I would like to acknowledge a special heartfelt debt to **Prof. E.C Nirmala Peter**, Professor in the department of Civil Engineering, JNTUH College of Engineering, Hyderabad for all she has done since I came from Iran in 2009 to start my M.tech study, for her genuine and sincere efforts in supervising the work, for her useful and practical suggestions and helpful discussions during the progress of the research. I was privileged to carry out my project under her guidance. I have learnt from her many things not only about this project but also about humanity and understanding people.

I would like to express my deepest thanks to **Prof. E. Saibaba Reddy** who has been extremely helpful throughout the study, for his guidance, patience, understanding and support, stimulating discussions and encouragement in these 2 years. He encouraged me to not only grow as an engineer but also as an instructor and an independent thinker. I am not sure many graduate students are given the opportunity to develop their own individuality and self-sufficiency by being allowed to work with such independence.

I am thankful to my beloved teachers **Prof. M. R. Madhav, Dr. V. Padamavathi, Dr. G.V. Narasimha Reddy, Dr. N. Darga kumar** and **Smt. M. Padmavathi**, who imparted me basic concepts of geotechnical engineering by their teaching and discussions.

Last but not least, my deepest gratitude goes to my beloved parents and my greatest teachers; **Mr. Ardavan Borzooei** and **Mrs. Masoumeh Rajab khorami** for their endless love, prayers and encouragement, for their faith in me and allowing me to be as ambitious as I wanted. It was under their watchful eyes that I gained so much drive and an ability to tackle challenges head on.

(SINA BORZOOEI)

ABSTRACT

Modeling of contaminant transport in porous media has been the focus of many researches in hydrology, soil science and environmental and geo-environmental engineering for many years. Researches mostly have been done from the 1950s and 1960s, created the basis of theoretical developments and analyses for the classical advection-dispersion equation (ADE). Classical advection-dispersion equation can be a satisfactory transport model in "homogeneous" porous media under some conditions, but It is well proven that such homogeneity rarely, if ever, exists. Wide range of heterogeneity of natural porous media necessitates development of more sophisticated transport theories.

The purpose of the present study is to compare analytical solution of ADE with two effective numerical and stochastic models based on finite difference and continuous time random walk to determine the most reliable and accurate method for contaminant transport modeling with emphasizing in transport in heterogeneous media. The final verification of proposed models with practical methods that accounts for the behavior of solute transport in heterogeneous media in laboratory with laboratory column tests data and published experimental data has been done too.

The Continuous time random walk model with truncated power law probability density function was developed by use of Matlab CTRW "toolbox" and Numerical modeling (E. C. Nirmala Peter 2007) considering power law mode of spatial variations of dispersion coefficient ($D_x = D_0 + mx^n$), has been developed and solved using the finite difference method for the one-dimensional flow and dispersion. The solutions obtained from the proposed models were verified and compared with the data observed in the laboratory experimental column tests, and also with the published experimental data of Wang et al. (1998) and long column experimental data of Huang et al. (1995).

Based on the results of comparison the breakthrough curves which was obtained from all methods, for reactive and non reactive contaminant, the best prediction was observed in CTRW model, while about numerical modeling in non reactive contaminant case the results were satisfactory but for reactive contaminant the numerical prediction could not be verified in full-fledged manner, and analytical solution of ADE model over-predicts the concentrations in break through curves.

Also a Comparison has been done to confirm the effect of using mean velocity (v_m) instead of seepage velocity in CTRW models, proposed by Ogata and Banks (1961), and far better

3

CTRW prediction was observed compare to the results of same model using seepage velocity.

With use of efficacious feature of CTRW toolbox for solving the inverse problems and fitting CTRW curves, dispersion coefficients were calculated for all collection points for experimental data, the results indicate that dispersion coefficients predicted by TPL and ADE(which is present in toolbox) and analytical (Fried (1975) methods are almost equal in shallow depth of sampling (below 30 cm), while the values will be diverged in more depth in such a way that TPL methods introduce greater value than ADE and analytical methods. Consequently the dispersion coefficient obtained simply by analytical solution can be valid just for the movement of contaminant in the shallow depth of soil.

The vast parametric Sampling based sensitivity analysis have been done for both proposed methods to clarify the effectiveness of each parameters in models .based on the results for CTRW truncated power law method exponent β observed as most effective parameters which can quantitatively describe the nature of solute transport (Fickian or non-Fickian).It was observed that with increasing the β value generally BTCs become sharper and less disperse and the nature of transport gradually changes from non-Fickian to Fickian.

According to the β values obtained from vast trial and error procedure, for each experimental data, it can usefully concluded that in most of the experiments the β values were between 1 and 2 which is representative of transition condition between fickian to non-fickian transport and with increasing the depth of sampling in column test, the gradual decline observed in β value, which can be construed in this way that with increasing the depth which means increasing the scale in one dimensional transport , the β value will decrease and the transport nature will change from fickian to non-fickian gradually or soil heterogeneities will increase by increasing the depth.

4

TABLE OF CONTENTS

LIST OF TABLES

8

9

LIST OF FIGURES

11

CHAPTER 1

INTRODUCTION

Geo-environmental Engineering is defined as "A field that encompasses the application of science and engineering principles to the analysis of the fate of contaminants on and in the ground, transfer of water, contaminant, and energy through geo-media; and design and implementation of methods for treating, modifying, reusing, or containing wastes on and in ground". A contaminant is generally defined as a miscible (aqueous) or immiscible (non-aqueous) phase liquid, or soluble or insoluble solid, which gets mixed with water as a consequence of human activities.

Soil is a dynamic system in which numerous chemical, physical and biological reactions occur singly or simultaneously. Evaluation of chemicals and fluid flows is an essential part of geo-environmental projects involving contaminant transport and waste containment. The concept of transport relating to the migration and fate of contaminants from waste sites, miscible in ground water is very much essential to prepare groundwater and contamination flow models. The transport of contaminants in soil involves important mechanisms such as molecular diffusion, dispersion under physical processes, adsorption, precipitation and oxidation-reduction under chemical processes and biodegradation under biological process.

Today's with enhancement in our knowledge about vast proven uncertainty in characterization of aquifer properties, groundwater movement and contaminant transport in naturally fractured and heterogeneous porous media has become a very complex matter and the main problem is how to model the stochastic and probabilistic uncertainties in porous media. Contaminant transport has been treated by using of the advection-dispersion equation (ADE), or by using some other related approaches, such as particle tracking random walk techniques which are basically based on the same assumptions as the ADE. These treatments include deterministic and stochastic approaches. But, as explained in many other researches results, these methods based on solution of ADE often fail to accurately model contaminant transport even in highly "homogeneous" systems. This failure can be justified and explained by the evidenced fact named as a "scale-dependent dispersion" which is in contrary to the fundamental assumptions of the ADE and declare that dispersivity is not constant term and it always varies as a function of time or distance traveled by the contaminant or both of them. This scale and time-dependent behavior of contaminant transport has created terms like "pre-asymptotic" or "non-Gaussian" ,"non-Fickian" or "anomalous" transport. An unusual early breakthrough times and long late time tails, in measured breakthrough curves can be the best

13

representative evidence for explaining this type of contaminant behavior. All of the accepted explanations for non-Fickian transport have this point in common that heterogeneities which cannot be ignored and present at all scales in media are the responsible and reason for this unusual behavior.

One approach to solve, better to say to simplify, this difficulty, is to apply a numerical code which incorporates the ADE using either partial differential equation, considering the hydraulic conductivity or velocity of field as a "functional form" which allows the dispersivity to change with travel distance or time which is obviously proven mathematically incorrect, can be easily found in contradict with the fundamental assumptions of ADE (Berkowitz and Scher, 1995).

An alternative, non-perturbative, probabilistic modeling framework which has been used successfully in quantification of non-Fickian transport in hydro-geological environment is made by CTRW theory (Berkowitz et al., 2004). The CTRW theory also has been used In particular, for capturing those mentioned unusual early and late time tailing observed in BTC measurements (Levy and Berkowitz 2003).

In the present study, for better understanding of the contaminant transport processes, physical modeling was carried out in the laboratory on different types of soils with non-reactive contaminants and reactive contaminants and the dispersion coefficients were obtained initially from the break-through curves using the classical Fickian type of CDE. It was established from these results that the dispersion coefficient is a function of distance of travel. Based on the result of previous work (E. C. Nirmala Peter 2007), considering power law mode of spatial variations of dispersion coefficient $(D_x = D_0 + mx^n)$, models have been developed and solved using the finite difference method for the one-dimensional flow and dispersion. On the other hand the CTRW modeling were developed by use of Matlab CTRW "toolbox" developed by Cortis and Berkowitz. The solutions obtained from the proposed models were verified and compared with the data observed in the laboratory column tests, and also for the published experimental data of Wang *et al.* (1998) and long column experimental data of Huang *et al.* (1995).

Review of literature with regard to laboratory studies carried out considering the scale-dependent dispersion and CTRW methods is given in Chapter 2. Chapter 3 presents the comprehensive background of CTRW methods and governing the Basic Formulation of Transport for non reactive and reactive contaminant and finally complete discussion about CTRW toolbox v3.1 mat lab code which used in this study.

The brief details of column tests in the laboratory are furnished in Chapter 4. Two types of experimental set-ups were used. The set-up with a maximum column length of 60 cm was provided with a single contaminant collection point, i.e. one opening at the bottom of the column. The second set-up with a maximum column length of 122 cm was provided with six contaminant sample collection points at various locations along the depth of the column. Break-through curves corresponding to each collection point location were presented in the same Chapter. The dispersion coefficients calculated corresponding to each contaminant collection point location is also presented in this Chapter. The results emphasized the variation of dispersion coefficient with depth. Numerical solutions of Convection Dispersion Equation (CDE), assuming variable dispersion coefficient as power law function with depth were presented in Chapter 4.

Chapter 5 furnishes the results of Sampling based sensitivity analysis for finite difference method and CTRW methods, all Break-through curves corresponding to sensitivity analysis for CTRW methods were presented in the same Chapter.

The comparison between CTRW models with use of seepage velocity and mean velocity for tests C_1 and C_2 were presented in chapter 6 to evaluate accuracy of Ogata and Banks (1961) results, In this chapter also the comparison between dispersion coefficients obtained from analytical models and CTRW truncated power law method and ADE method with use of mean velocity were presented, finally for all the tests in addition to experimental data of Huang et al. (1995) and WANG et al. (1998) verification of CTRW and Finite difference models were done and all the input parameters were tabulated in this chapter , comparison of Break-through curves corresponding to CTRW, ADE and finite difference methods with experimental data were presented in the same Chapter.

CHAPTER 2

REVIEW OF LITERATURE

2.1 Introduction

Any plan of mitigation, cleanup operations, or control measures, once contamination has been detected in the subsurface, requires the prediction of pathways and fate of the contaminants, in response to certain planned remediation activities. Similarly, any monitoring or observation network must be based on the anticipated behaviour of the system. Management means making decisions to achieve goals, without violating specified constraints. Therefore, good management requires information on the response of the managed system to proposed activities. This information enables the planner, or the decision-maker, to compare alternative actions, to select the best one, and to ensure that constraints are not violated. All such predictions can be obtained, within the framework of a considered management problem, by constructing models of the investigated domain, and of the flow and solute transport phenomena that take place in it. The model can be physical (for example, a laboratory 'sand-box' and field lysimeter), electrical analogies, 'black-box' statistical models and mathematical models, which involve both analytical and numerical techniques.

2.2 Ground water modeling approaches:

The large number of contaminated groundwater supplies highlights the need for tools to evaluate the subsurface processes controlling contaminant transport. In the evaluation of contaminated sites it is not uncommon that the depth of the vadose zone is not specifically represented, and that the permeability of the whole aquifer is represented by a single hydraulic conductivity value. In order to improve the evaluation and prediction of contaminant movement, and provide decision support for remediation strategies, it is essential that contaminant transport is calculated in more detail. Specifically, it is crucial that the transit time from the source to sensitive recipients is determined more precisely, and that the concentration of contaminant by the time it reaches the recipients is predicted more accurately. Because the governing equations for subsurface flow and mass transport are partial differential equations (PDE), it is necessary to define appropriate boundary conditions. A direct analytical solution is only possible for domains of simple geometric shape or for homogeneous media (Strack, 1999). Instead of trying to solve the PDE, one alternative option is to use a simple non distributed mass-balance approach, based on the contaminant mass and the water balance, to calculate the dilution of contaminant leakage into the aquifer . However, the usual method is to use a distributed model, in which the study area is subdivided into small, finite, calculation cells, and to replace the PDE by approximate, algebraic functions. The main numerical approaches used in practice are the finite-difference method (FDM) and the finite-element method (FEM). However, FEM has not been used frequently in earlier groundwater modelling exercises, probably because it is more demanding of computing power. Subsurface water flow is normally independent of the contaminant being transported. The exception is when the concentration of contaminants affects the water density and hence the water flow. For water soluble, moderately concentrated contaminants, this means that there is no Requirement to solve the flow and transport equations simultaneously (Domenico & Schwartz, 1998). Therefore, a common model approach is first to calculate the velocity field, and second to calculate the mass transport. This approach also has the advantage that it can generate stable, converging solutions of the flow field, and that it is possible to include more sophisticated geochemical models.

Groundwater flow is a simplified case of flow in the vadose zone. It is therefore common to use different modelling approaches, each optimized for each of the zones. Several authors have proposed coupled models for transport in the vadose and groundwater zones (Panday & Huyakorn, 2008; Twarakavi et al., 2008), but models also exist that operate for both zones simultaneously (Hughes & Liu, 2008; Trefry & Muffels, 2007). In general, distributed numerical models vary in complexity and the most comprehensive ones take into account the heterogeneity of the modelled environment.   Their practical application is, however, often hindered because there is insufficient data that justify their complexity (Hantush et al., 2000). Nevertheless, at specific sites with adequate field data, it is often appropriate to perform more detailed calculations through simulations in a distributed numerical model. Other options include Monte Carlo simulations based on parameter distribution functions (Soutter & Musy, 1997), the closely related fuzzy-set theory (Zhang et al., 2009; Li et al., 2007), and other methods that have been suggested as being more specifically capable of handling prediction uncertainty (Li et al., 2003; Wong & Yeh, 2002).Depending on the extent to which a site-specific model has been validated, it can be used to predict the behaviour of contaminants over reasonably long time-scales and under a variety of conditions .

Fewer of these models have been validated with data on groundwater contamination. Nevertheless, the combination of independent data types would improve the site-specific knowledge of flows and contaminant movements. Many studies have related hydro geological models to the remediation of contaminated sites. Several authors have either considered contaminant transport in groundwater at a regional scale (Szucs et al., 2009; Maqsood et al., 2005; Zhou et al., 2004; Schäfer & Therrien, 1995) or if both the vadose and groundwater zones have been included (Stenemo et al., 2005; Jorgensen et al., 2004), the studies have been carried out at a local scale.

The numerical problem for predicting contaminant transport in the vadose zone and in groundwater often becomes extremely demanding of computational power. A review of the literature shows that there has been an increasing tendency for the numerical problems to be solved on networks of computers, which are not publically available. One example is a study on the effectiveness of aquifer remediation (Maxwell et al., 2008), in which the numerical problem was solved on a network of 200 processors. Another fact that emerges from the literature is that synthetic examples dominate and, if field sites with real data are considered, the use of hypothetical contaminants is common (Maji & Sudicky, 2008; Maxwell et al., 2008; Rauber et al., 1998).

2.2.1 Deterministic approach

Deterministic models are based on fundamental notions of mathematical physics on hydro geological processes with synonymously defined causes and their consequences. They consider the movement (in general cases unsteady and spatial) of groundwater with solutes in pores and fissures of geological formations. Movement (mass transport) of compounds is advective and diffusive (dispersive). Advection is the transport of compound on the macro level of compounds (water and salt transport within pore solutions). Diffusion implies the transport on a molecular level due to movement of micro particles.

All compounds undergo the physical and chemical transformations, which occur permanently. Close to the ground level, there is an aeration zone composed of unsaturated geological material, which is highly aerated and contains water vapours due to the proximity of above surface level conditions. Under the aeration zone, in the saturated zone, the groundwater (layer) is situated. Deeper, there are groundwater layers with active water exchange and impacts of human activity that are the focus of study.

This brief description on the status of solutes of geological material shows how complicated the geological processes are. However, studying the definite territories, in a majority of cases we deal with temporal and spatial effects of a certain set of natural or designed conditions. These allow the study to be conducted with accuracy, sufficient for the practical application of obtained results. Integrated equations of mass transport consisting of

18

equations of mass balance, laws of movement and laws of unsteady thermodynamics are verified and presented in many publications (Cherny, 1963; Lukner and Shestakov, 1986; Ognianik et al., 1985,1977; etc.).

Integrated equations of mass transport are based on assumptions of a representative elementary volume, such as:

- geometrical homogeneity and averaged properties of porous geological material;
- known forces affecting the liquid;
- Known properties of liquid itself and velocities of its flow during the time periods, etc.

These assumptions allow us to perform the approximations of dissipation. The problem is to find out if the errors of these approximations will satisfy the practical objectives of mass transport.

2.2.2 Stochastic approach

Stochastic models of hydrological processes may be classified as follows:

1. Deterministic boundary tasks of filtration (mass transport) and, on the whole, probabilistic solution of the problem.

2. Deterministic mathematical description of the process (differential equations and boundary conditions), probabilistic description of medium properties in the studied area with randomized and statistic distributions.

3. statistic boundary conditions (hydrological regime of boundary water bodies and watercourses, hydro physical and hydro chemical conditions at the soil surface, etc.) providing that deterministic mathematical description of the process itself is within the area.

4. Probabilistic and probability statistical models.

The first group of models may be computed by Monte-Carlo techniques. The second and third group of models is typical for the majority of models for real fields of geo filtration. This is due to the exact value of their parameters within their probable variability that simulate, with a certain probability, the potential functions and velocities of mass transport. If data are available, the simulation models are applied; the influence of some parameters on the process is estimated and acceptable errors are established. They serve as a base for planning and implementation of explorations and experimental investigations for the creation of a model of appropriate probability.

The fourth group of models is known as "black box" and probability statistical models. Sometimes, it is helpful to treat a pollution event as a "black box". With regard to the entrance and exit of pollution, the water layer is considered as the 'black box. The mechanism inside the "black box" is unknown. The effect of the "black box" is examined by the incoming and outgoing functions of pollution. This method can be applied on a large scale, when pollution of the studied area is multi-component, but can be integrated in one image. Application of probability and statistical methods do not establish functional relationships but rather correlation between fluctuation of potential function (level, migrant concentrations) and one or few factors of the regime.

These methods are applied when there are long term observation data for the subject of prognosis or subject analogy and the position of potential functions may be extrapolated. Usually, by this method, the average annual concentrations of migrants, water levels of the current year or number of following years, or certain date or period are calculated. The paired and multiple correlated relations, regression models and harmonic analysis are usually applied and they are well described in existing publications (Kisle, 1972; Zaltzberg, 1976; Devis, 1977, etc.). The cybernetic method of mathematical modelling, namely the method of group assemblage of augments (Ivakhnenko, 1975), should be mentioned as well. This method is based on the principle of heuristic self organisation.

Most computer models utilize a deterministic approach where all data are input as single, "best estimate" values. Single value inputs result in single value outputs. When modelling on a site-specific scale, where extensive data has been collected and spatial characterization is well established, a deterministic approach is generally appropriate. Simulations with appropriate calibration, sensitivity analysis, and history matching can produce an adequate representation of the real hydrogeology system. If the modeling effort utilizes very limited data or where a larger, regional scale is involved, a stochastic (statistical) approach may be acceptable (e.g., Monte Carlo simulations).

This approach utilizes hydraulic parameters having a probability distribution that results in all output having the same probability distribution. A stochastic approach to modeling would characterize parameter uncertainty by incorporating a measure of uncertainty into the parameters and database utilized in the simulations. When a lack of data and a high degree of data uncertainty exists, calibration and additional history matching can be long, tedious, or impossible.

The stochastic approach allows the uncertainty factor to be maintained throughout the modeling process, allowing for potentially more realistic interpretations of the results by providing ranges of scenarios applicable to the real system. Too often, the data uncertainty factor is lost when deterministic approaches are utilized at sites for which limited data are available. The results become "fact" without acknowledgment of the limitations dictated by the input parameters and the underlying assumptions

2.2.3 Analytical and numerical approach

The governing equations for groundwater systems are usually solved either analytically or numerically. Analytical models contain analytical solution of the field equations, continuously in space and time. In numerical models, a discrete solution is obtained in both the space and time domains by using numerical approximations of the governing partial differential equation. Various numerical solution techniques are used in groundwater models. They include Finite-Difference Methods, integral Finite-Difference methods, Galerkin and variable Finite-Element Methods, Collation Methods, Boundary-Element Methods, Particle Mass Tracking Methods (e.g. Random Walk), and Methods of Characteristics. Among the most used approaches in groundwater modelling, three Techniques can be distinguished:
• The Finite Difference Method (FDM),
• The Finite Element Method (FEM) and
• The Analytical Element Method (AEM).

All techniques have their own advantages and disadvantages with respect to availability, costs, user friendliness, applicability, and required knowledge of the user, all of which may vary on both sides of a national border.

The Finite Difference Method is based on quadrangular grids that apply over the vertical. The aquifer system is discretisised into a mesh of points termed 'nodes', forming rows, columns and layers. Conceptually, the nodes represent prisms of porous material, termed cells, in which the hydraulic properties are constant and so that any value associated with a node applies to or is distributed over the extent of a cell. For defining the configuration of cells with respect to the location of nodes, two conventions exist: the 'block centered' and the 'point-centered' formulations. Both systems start by dividing the aquifer with two sets of parallel lines, which are perpendicular to each other.

In the block-centered formulation, the blocks formed by the sets of parallel lines are the cells; the nodes are at the centre of the cells. In the point-centered formulation, the nodes are at the intersection points of the sets of parallel lines, and cells are drawn around the nodes with faces halfway between nodes. In both cases, spacing of nodes should be in such a way that the hydraulic properties of the system are, in fact, uniform over the extent of a cell. The Finite Element Method is generally based on triangular grids but can also be based on quadrangular grids. The areas of the elements can vary by gradually changing the node spacing (Figure 2.1). (Hemker, 1997).

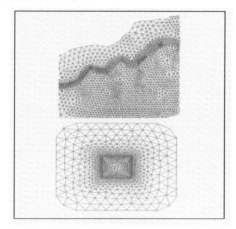

Figure 2.1 Examples of finite element grid generation (meandering ribvers and sheet piling or excavations) (Hemker, 1997)

The finite element method solves the differential equation for semi-three dimensional groundwater flow per element, using a minimization of a certain error all over the model In general; finite element grids are more flexible than finite difference grids for simulation of special boundaries such as river courses. The major application aspects are not much different from that of finite difference methods.

The Analytical Element Method is based on analytic solutions of the differential equation describing groundwater flows (De Lange, 1991). A model is constructed by combining different types of analytic elements, which are based on the principle of superposition. Examples of analytic elements are the well, the line-sink, the area-sink and the in homogeneity.

Each type of analytic element can simulate different types of geo hydrological features (abstraction wells, rivers, polders, infiltration areas, aquifer in homogeneities, leakage layers between aquifers etc.). Actually, modeling with analytic elements is the modeling of geo hydrological features rather than generating parameter values of elements in a mesh, like in FDM and FEM. Analytical elements apply in infinite aquifers. Therefore, a model of analytic elements is not bounded as a model based on finite element or finite difference modeling techniques. The boundary of a model of an analytic element is actually a surrounding zone that also consists of analytic elements generating the effects of the outside world on the actual modeling domain (the domain of interest).

Interchanging elements of the surrounding zone and those of the connected model carry out connection of models. The Finite Difference Method and the Finite Element Method are grid-based (quadrangular or triangular respectively). Because of the shape of the elements, the FDM with quadrangles has less flexibility in schematization than FEM with triangles.

The Analytic Element Method is the most flexible with respect to schematization. Any form or size of the element is possible relating to the area of interest.

For the FDM and FEM where this is not always the case, the accuracy largely depends on the iteration criteria. With the use of post processors, this problem can be corrected.

When using the Finite Difference Method or the Finite Element Method, many data are needed because all cells need to be filled.

2.3 Ground water contaminant transport modeling

Decades of researches have focused on how to quantify contaminant transport in porous soils and rock. Landmark experiments dating mostly from the 1950s and 1960s formed the basis for theoretical developments and analysis, which considered almost exclusively the classical ADE. In the usual one-dimensional·form, the ADE is simply:

$$\frac{\partial}{\partial t} C(x,t) = -v \frac{\partial C(x,t)}{\partial t} + D \frac{\partial^2}{\partial x^2} C(x,t) \qquad (2.1)$$

Where c is the resident fluid concentration, t is time, x is distance, and D is the hydrodynamic dispersion coefficient. The average fluid velocity v is equal to q/n, where q is the volumetric flux per unit area of porous medium, and n is porosity.

The ADE concept assumes that velocity and dispersivity coefficients will remain constant during the transport procedure and also complete homogeneity of the porous medium, at least at the relevant scale of measurement in other words, as clarified by many researchers like Bear (1972), Berkowitz and Scher (2001) and Berkowitz et al. (2002), the equation is founded on the fundamental assumption that dispersive transport follows a Fickian behavior.

However, results of previous researches in this area (Hoffman et al., 1996; Oswald et al., 1997) have been proven the existence of phenomena, as they called, preferential flow paths due to macro-structure and microstructure heterogeneities of the porous media which can strongly affect both water flow and tracer transport in carefully packed laboratory-scale flow cells and columns, by using very accurate methods like magnetic resonance imaging to visualize flow conditions within "homogeneous" geologic materials in laboratory-scale column experiments .

Furthermore some of the other researchers like Silliman and Simpson (1987) have shown even in the laboratory experiment, the scale dependency of the dispersivity coefficient can be proven easily which is completely in contrast to the fundamental assumption of ADE.

Consequently about 1950 a series of systematic errors in fitting BTCs using the classical ADE were observed in many researchers' works. Aronofsky and Heller in 1957 and Scheidegger in 1959, proved that these systematic errors in column test cannot be easily explained by usual variability in experimental measurements so finally they concluded the diffusivity equation up to this error in tracer experiments.

The main difference between BTCs predicted by the ADE and BTCs obtained considering the heterogeneity effect of porous media was reported as anomalous early arrival times and long late time tails and having delay in both early and late times. This anomalous behaviour in BTCs and transport of tracer made some of researchers to name it as anomalous or non-Fickian, pre-asymptotic or pre-ergodic behaviour (Levy and Berkowitz, 2003).

Considering mentioned problem, recently many efforts have been done to develop model for anomalous contaminant transport, in Dagan and Neuman, 1997; Dagan, 1989; Gelhar, 1993 stochastic perturbative approaches based on an ensemble-averaged ADE, have been used for modelling transport in mildly heterogeneous porous media.

In some studies Perturbative methods have been applied for the problematic non-Fickian transport behaviours observed in field experiments (e.g., Boggs et al., 1992; Adams and Gelhar, 1992) and also (e.g., Burr and Sudicky, 1994; Naff et al., 1998; Salandin and Fiorotto, 1998; Pannone and Kitanidis, 2001; McLaughlin and Ruan, 2001; Dentz et al., 2002) have applied numerical transport simulations. However, the absence of Perturbative methods in laboratory-scale flow systems can be felt in reviewing previous researches.

2.3.1 Continuous time random walk method

The first introduction of Continuous Time Random Walk (CTRW) method was by (Montroll and Weiss, 1965). Before that the method was used to describe systems out of equilibrium, like anomalous diffusion phenomena which can be characterized by non– linear time dependence of mean squared displacement (MSD). The CTRW framework can account for a very wide range of non-Fickian and Fickian transport behaviours, with a smooth transition from non-Fickian to Fickian behaviour.

In a simplest way, CTRW method can be describe as a jumping model which considers the diffusion of a particle as a sequence of independent random instantaneous transitions or jumps. The movement of particles can be completely quantified by two probability density functions for transition length and waiting – time between the successive jumps as independent random variables.

The first application of CTRW theory for modelling the transport in fracture networks (which was applied to numerical studies) was introduced by Berkowitz and Scher in 1998 and for studying macro-dispersion experiment or tracer test (Berkowitz and Scher, 1998).

As an very important fact, Berkowitz and Scher (1995,2001) for the first time introduced correlation between characteristic time scales (power-law transport time distributions) with distribution of heterogeneity length scales .

As an advantage, CTRW framework can be validated for special cases relevant to different forms of the ADE, and variety of mobile-immobile and multirate mass transfer models can be derived as special cases within the CTRW framework (e.g., Berkowitz et al., 2002; Dentz et al., 2004).

Fractional-in-time and in-space derivative formulations for transport equation are two different forms of CTRW which have been studied in different researches. (Metzler and Klafter, 2000; Pachepsky et al., 2000; Berkowitz and Scher, 2001; Berkowitz et al., 2002; Scher et al., 2002; Schumer et al., 2003). However many researches have been done about their limitations and deficiencies for their applications in contaminant transport in soil (Lu et al., 2002; Zhou and Selim, 2003).

As a very seminal advantage, in contrary to these two types of Fractional derivative formulations, CTRW model for contaminant transport can cover continuously from non-Fickian type of transport due to high range of heterogeneities to Fickian type which is caused under highly restrictive conditions by high degree of homogeneity in the hydraulic conductivity over long spatial and temporal scales. (Berkowitz and Scher (2000), Berkowitz et al. (2002), and Cortis et al. (2004).

2.3.2 Numerical models of contaminant transport

Since most analytical techniques treat only ideal homogenous and isotropic porous medium and they are not directly applicable to most field situations, numerical simulation must be considered. In a no homogeneous and anisotropic groundwater flow system, the groundwater seepage velocity is not constant. The character of the advection-dispersion may vary in space and time depending on the velocity field. The advection-dispersion equation becomes a parabolic type partial differential operation if the Peclet number is small. The Peclet number is physically interpreted as a ratio of advective to dispersive transport components. The equation becomes a nearly hyperbolic type partial differential equation if the Peclet number is large, i.e., transport-dominated system. Early numerical experiments for solving the advection-dispersion equation were based on "straightforward" finite differences (Peaceman and Rachford, 1962; Stone and Brian, 1963; Shamir and Harleman, 1967). Finite differences perform well in dispersion-dominated situations where the Peclet number is small. In a transport-dominated flow system the concentration gradient is usually very steep. Difficulties arise, such as numerical dispersion and oscillations, in the numerical simulation of this sharp front. Many numerical methods have been developed in the last two decades to handle this sharp front. Common numerical approaches for the advection-diffusion equation include the implicit diffusive method, Galerkin finite element and collection methods, alternating direction method, as well as the method of characteristics.

27

2.4 Present study

The Convection-Dispersion Equation (CDE) model has been quite successful in describing results from laboratory studies involving carefully constructed homogeneous columns. In the classical CDE, the dispersion coefficient is assumed constant, but the results from several field studies (Taylor and Howard (1987), Domenico and Robbins (1984), Toride et al., 1995) indicated scale-dependent dispersion i.e. dispersion coefficient increases with distance 'x' from the source of pollution. Many analytical solutions have been developed by Mishra and Parker (1990), Khan and Jury (1990), Pachepsky et al. (2000), Pang and Hunt (2001)). These solutions are difficult to implement due to their highly idealized boundary conditions and parameters. As an alternative, numerical methods are often preferred as these methods are easier to implement than the analytical solutions. In the present work numerical model was developed considering power law variations of dispersion coefficient with distance 'x' from the contaminant source for both non-reactive and reactive contaminants.

The laboratory experimentation involved studying the presence of scaling relationships of transport property i.e. dispersion coefficient with the length of the column by conducting column tests with data collection points at different depths / locations along the length of the column. Considering power law mode of spatial variations of dispersion coefficient ($D_x = D_0 + mx^n$), models have been developed and solved using the finite difference method for the one-dimensional flow and dispersion. On the other hand the CTRW modelling were developed by use of Matlab CTRW "toolbox" developed by Cortis and Berkowitz. The solutions obtained from the proposed models were verified and compared with the data observed in the laboratory column tests, and also for the published experimental data of Wang et al. (1998) and long column experimental data of Huang et al. (1995).

CHAPTER 3

CONTINUOUS TIME RANDOM WALK MODELING OF ONE-DIMENSIONAL CONTAMINANT TRANSPORT

3.1 INTRODUCTION

Decades of researches have focused on how to quantify contaminant transport in porous soils and rock. Landmark experiments dating mostly from the 1950s and 1960s formed the basis for theoretical developments and analysis, which considered almost exclusively the classical ADE.

The ADE concept assumes that velocity and dispersivity coefficients will remain constant during the transport procedure and also complete homogeneity of the porous medium, at least at the relevant scale of measurement in other words, as clarified by many researchers like Bear (1972), Berkowitz and Scher (2001) and Berkowitz et al. (2002), the equation is founded on the fundamental assumption that dispersive transport follows a Fickian behavior, which means mechanical dispersion can be quantified by a macroscopic type of Fick's law and furthermore the temporal and spatial concentration distributions of test results follow normal or Gaussian distribution. (Berkowitz et al. 2006)

However, results of previous researches in this area (Hoffman et al., 1996; Oswald et al., 1997) have been proven the existence of phenomena, as they called, preferential flow paths due to macro-structure and microstructure heterogeneities of the porous media which can strongly affect both water flow and tracer transport in carefully packed laboratory-scale flow cells and columns, by using very accurate methods like magnetic resonance imaging to visualize flow conditions within "homogeneous" geologic materials in laboratory-scale column experiments . The heterogeneity of porous media has been proven highly variable in scale and all scales of heterogeneity, vast range of velocities and stagnation areas in porous media and any other changes in biogeochemical properties of porous media which can cause temporarily or permanently time shifting in contaminant transport (mostly in diffusion-dominated regions) have been shown tangibly effective in contaminant transport. (Berkowitz et al. 2006)

Furthermore some of the other researchers like Silliman and Simpson (1987) have shown even in the laboratory experiment, the scale dependency of the dispersivity coefficient can be proven easily which is completely in contrast to the fundamental assumption of ADE.

29

Overall, about 1950 series of systematic errors in fitting BTCs using the classical ADE were observed in many researchers' works. Aronofsky and Heller in 1957 and Scheidegger in 1959, proved that these systematic errors in column tests cannot be easily explained by usual variability in experimental measurements so finally they concluded the diffusivity equation up to this error in tracer experiments.

The main difference between BTCs predicted by the ADE and BTCs obtained considering the heterogeneity effect of porous media was reported as anomalous early arrival times and long late time tails and having delay in both early and late times. This anomalous behaviour in BTCs and transport of tracer made some of researchers to name it as anomalous or non-Fickian, pre-asymptotic or pre-ergodic behaviour.

This concept was first studied and introduced by (Montroll and Scher,1973) and (Scher and Montroll,1975) and it has been proven so applicable to transport and diffusion in disordered systems (Levy and Berkowitz, 2003).

Considering mentioned problem, recently many efforts have been done to develop model for anomalous contaminant transport, in (Dagan and Neuman, 1997);(Dagan, 1989); (Gelhar, 1993) stochastic perturbative approaches based on an ensemble-averaged ADE, have been used for modelling transport in mildly heterogeneous porous media.

In some studies Perturbative methods have been applied for the problematic non-Fickian transport behaviours observed in field experiments (e.g., Boggs et al., 1992; Adams and Gelhar, 1992) and also (e.g., Burr and Sudicky, 1994; Naff et al., 1998; Salandin and Fiorotto, 1998; Pannone and Kitanidis, 2001; McLaughlin and Ruan, 2001; Dentz et al., 2002) have applied numerical transport simulations. However, the absence of Perturbative methods in laboratory-scale flow systems can be felt in reviewing previous researches.

The purpose of this chapter is to introduce and explain more details about an effective theoretical method for modeling the behavior of solute transport in heterogeneous media in field, laboratory, and numerical experiments based on the continuous time random walk (CTRW) as an non-perturbative method.

The first introduction of Continuous Time Random Walk (CTRW) method was by (Montroll and Weiss, 1965). Before that the method was used to describe systems out of equilibrium, like anomalous diffusion phenomena which can be characterized by non– linear time dependence of mean squared displacement (MSD). The CTRW framework can account for a very wide range of non-Fickian and Fickian transport behaviours, with a smooth transition from non-Fickian to Fickian behaviour.

30

As an very important fact, Berkowitz and Scher (1995,2001) for the first time introduced correlation between characteristic time scales (power-law transport time distributions) with distribution of heterogeneity length scales .

Fractional-in-time and in-space derivative formulations for transport equation are two different forms of CTRW which have been studied in different researches. (Metzler and Klafter, 2000; Pachepsky et al., 2000; Berkowitz and Scher, 2001; Berkowitz et al., 2002; Scher et al., 2002; Schumer et al., 2003). However many studies have been done about their limitations and deficiencies for their applications in contaminant transport in soil (Lu et al., 2002; Zhou and Selim, 2003).

As a very seminal advantage, in contrary to these two types of Fractional derivative formulations, CTRW model for contaminant transport can cover continuously from non-Fickian type of transport due to high range of heterogeneities to Fickian type which is caused under highly restrictive conditions by high degree of homogeneity in the hydraulic conductivity over long spatial and temporal scales (Berkowitz and Scher (2000), Berkowitz et al. (2002), and Cortis et al. (2004)).

In this thesis the CTRW modeling was developed by use of Matlab CTRW ''toolbox'' developed by Cortis and Berkowitz (Cortis, A., and B. Berkowitz, 2005).

3.2. The Conceptual Picture of Continuous time random walk

Contaminant movement in any geological media can be conceptualized by considering particles undergoing various transitions including displacement due to heterogeneity and stagnation and time taken movements. These transitions in CTRW theory can cover all changes in contaminant transport regime due to hydraulic and geochemical, macro and micro heterogeneities of the field.In a simplest way, CTRW method can be describe as a jumping model which considers the diffusion of a particle as a sequence of independent random instantaneous transitions or jumps. The movement of particles can be completely quantified by two probability density functions for transition length and waiting – time between the successive jumps as independent random variables.

Each of these transitions can be treated as w(s, s') in quantifiable manner, which is the rate of particle transfer from position s' to s. In this type of interpretation of particle movement, particles or chemical species move with different velocities via different paths and the effects of multiscale heterogeneities on these transition is impregnable, on the other hand the magnitude of particles transition times and length are not necessarily correlated, consequently for considering the vast statistical variation in porous media properties, CTRW model utilizes a coupled probability density function (pdf), $\psi(s,t)$, which can describe particles transition for given distances, in given directions, and in given times. (Berkowitz et al. 2006).

With studying the functional shape of the pdf, CTRW framework can Identify the nature of contaminant transport, non-Fickian or Fickian and also can be validated for special cases relevant to different forms of the ADE, and variety of mobile-immobile and multirate mass transfer models can be derived as special cases within the CTRW framework (e.g., Berkowitz et al., 2002; Dentz et al., 2004).

3.3. Basic Formulation of Transport for non reactive contaminant

The Equation(3.1) based on a conservation of mass theory, which was known as the "master equation" (ME) first was introduced by Oppenheim *et al* in 1977 and Shlesinger in 1996 , widely used afterwards in the physics and chemistry and electron movement in random semiconductors (Klafter and Silbey, 1980a)

$$\frac{\partial C(s,t)}{\partial t} = -\sum_{s'} w(s',s)C(s,t) + \sum_{s'} w(s,s')C(s',t) \quad (3.1)$$

Where C(s, t) is the probability at point s and time t in a specific realization of the domain and the dimension of $\Sigma_s w$ is reciprocal time and w(s, s') is the rate of particle transfer from position s' to s. The master equation in contaminant transport is not able to differentiate an advective and dispersive part of the motion. (Berkowitz et al. 2006)

In 1980 Klafter and Silbey proposed generalized master equation(GME) which was the ensemble average of (3.1) and nonlocal in time with usage of ensemble-averaged, normalized concentration and integral over time requiring knowledge of the past state of the concentration to improve the probabilistic approach in (ME).

Afterwards it was proven that with use of the Laplace transform, GME equation can be converted to CTRW formulation (Kenkre et al., 1973; Shlesinger, 1974). The heart of the CTRW formulation was probability density function of particles transitions known as $\psi(s,t)$. This function can describe each particle transition over a distance and direction s in time t and it was proven as the best indicator of effect of properties of porous media, degree of water saturation, fracture matrix and adsorption or desorption and in short term it is related to a wide range of mechanisms which can influence on particles transitions (Margolin et al., 2003).

Based on the assumption of existence of spatial moments of pdf, the function can be decoupled in to $p(s) \psi(t)$, where $\psi(t)$ is the probability rate for a transition time t between sites and p(s) is probability distribution of jump lengths. P(s) function has been assumed to have finite first and second moments which can define coefficients of the transport velocity and dispersion, however in any other way it is a general positive function with a long tail which can be considered as a Gaussian distribution after a large number of steps for the numerical simulations (Berkowitz et al. 2006).

Using the decoupled form of probability density function, (Berkowitz et al., 2002; Dentz et al., 2004; Cortis et al., 2004b) the special type of Fokker-Planck equation with considering continuous time random walk over assumed statistical homogeneities scales has been proposed in Laplace space.

In one dimension, the Laplace transformed concentration $\tilde{C}(x,u)$ is given by:

$$u\tilde{C}(x,u) - C_0(x) = -\tilde{M}(u)\left[v_\psi\frac{\partial}{\partial x}\tilde{C}(x,u) - D_\psi\frac{\partial^2}{\partial x^2}\tilde{C}(x,u)\right] \quad (3.2)$$

Where

$$\tilde{M}(u) = \bar{t}u\frac{\tilde{\psi}(u)}{1-\tilde{\psi}(u)} \quad (3.3)$$

$$v_\psi = \frac{1}{\bar{t}}\int p(s)s\,d^ds \quad (3.4)$$

$$D_\psi = \frac{1}{\bar{t}}\int\frac{1}{2}p(s)ss\,d^ds \quad (3.5)$$

Where, u is the dimensional Laplace variable with units of $[t^{-1}]$ and the $\sim$ symbol represents the Laplace transformed variable. $\tilde{M}(u)$ is a memory function which accounts for the undefined small-scale heterogeneities and can take on several expressions, depending on the functional form of $\tilde{\psi}(u)$ which is the heart of the CTRW formulation, and characterizes the nature of the solute movement. $\bar{t}$ is a characteristic (median) time. v_ψ and D_ψ are the transport velocity and generalized dispersion coefficient respectively, which have different physical interpretations than average fluid velocity and dispersion in the usual ADE definition.

3.4 Models of $\tilde{\psi}(u)$ (Heart of CTRW methods)

As it was mentioned $\tilde{\psi}(u)$ is the heart of CTRW methods, which can be used in advective, diffusive, dispersive transport modeling in over a wide range of timescales or for transport of particles in fully and partially saturated media in domains containing flowing and stagnant zones or for modeling the reactive contaminant which sorption has an important effect in them. In following sections the main and the most common models of $\tilde{\psi}(u)$ will be discussed however it is proven that any model of $\tilde{\psi}(u)$ should be chosen by considering the physical nature of the system. Generally four types of $\tilde{\psi}(u)$, truncated power law, asymptotic and the modified exponential and Fractional-in-time derivative equation forms are used.

3.4.1. Truncated Power Law

Previous researches proved the importance of large power law region in making the practical transport model, however any real physical system has been shown that transport tends to become normal or Fickian at large enough time (cut-off timescale) if the scale of observation is larger than the largest heterogeneity scale.

To bring this practical model to mathematical formulation, (3.6) equation has been presented the $\psi(t)$ by a truncated power law distribution which can cover transitions from anomalous to normal transport by use of this cut-off behavior (Dentz et al., 2004).

$$\psi(t) = \left[t_1 \tau_2^{-\beta} \exp(\tau_2^{-1}) \, \Gamma(-\beta, \tau_2^{-1}) \right]^{-1} \frac{\exp(1+t/t_2)}{(1+t/t_1)^{1+\beta}} \qquad (3.6)$$

and the Laplace transform of pdf is given by

$$\psi(t) = (1 + \tau_2 u t_1)^{\beta} \exp(t_1 u) . \frac{\Gamma(-\beta, \tau_2^{-1} + t_1 u)}{\Gamma(-\beta, \tau_2^{-1})} \qquad (3.7)$$

Where $\tau_2 \equiv t_2/t_1$ and $\Gamma(a, x)$ is the incomplete Gamma function ,the time t_1 (approximate median transition time) is the lower limit from which power law behavior starts and the time t_2 as a cut-off time is an upper limit which after that transport evolves into a normal condition and the exponent β is a very useful means to characterize the latter range. In fact β is a function of the length scale or time needed to traverse this length scale which can be influenced by the characteristic length of the largest scale heterogeneity, for validating the C.T.R.W method, β should be constant or slowly varying over a number of orders of magnitude in length of time. Consequently, it can be concluded that truncated power law will be characterized by the exponent β and by the two timescales t_1 and t_2 which are indicators of duration of the observation and the extent of disorder in this model.

For transition time $t_1 \ll t \ll t_2$, the nature of contaminant transport can be characterized from Fickian to non-Fickian by exponent β and its magnitude thereby, the transport behavior is anomalous for $0<\beta<2$ While for t $>>t_2$, pdf decreases exponentially and the transport behavior becomes Fickian. For $0 < \beta <1$ and realistically for $\beta >0.2$ the pure anomalous behaviour will be observed with asymmetrical BTCs with delayed early and long late time tails. For $1 <\beta< 2$ as transmission condition from anomalous to Fickian transport, BTCs still are asymmetric with long late time tails but sharper and less disperse. Compared to previous condition the mean location of the plume moves with a constant velocity and the standard deviation of that scales as t $^{(3-\beta)/2}$ instead of t$^{\beta}$. It is notable to set the cut-off time t_2 to a very large value to have a pure power law model. Finally for $\beta >2$, the contaminant transport model tends to behave as the classical Fickian described by the ADE model. In this form the importance of upper cutoff t_2 time decreases on the transport behaviour. Mean location of the plume moves with the average velocity and BTCs are in simple Fickian shape and the standard deviation scales as $t^{1/2}$ (Shlesinger, 1974; Dentz et al., 2004; Berkowitz et al., 2006).

3.4.2. Asymptotic Behaviors

Asymptotic behavior referred to particle transition behavior which is in sharp contrast to Gaussian models, considering the non locality using of time dependent coefficient in pde is not the conceptually correct choice. This model was proposed in a related from to the general equation of CTRW $\widetilde{\psi}\,(t) \propto \tau^{-1-\beta}$ by Cortis et al in 2004.

$$\widetilde{\psi}\,(t) = \left(1 + au + bu^{\beta}\right)^{-1} \quad (3.8)$$

3.4.3. The modified exponential or ETA model

$$\tilde{\psi}(u) \equiv \mathcal{L}\left\{2\eta_3 F_3 \begin{bmatrix} 1,1,1 \\ 2,2,2 \end{bmatrix}; -\tau \right] e^{-2\eta\tau_4 F_4 \begin{bmatrix} 1,1,1,1 \\ 2,2,2,2 \end{bmatrix}; -\tau \right]}; \tau \to u \right\}, u > 0 \quad (3.9)$$

Where η is the disorder parameter and $\mathcal{L}$ is the Laplace transform. Hereby η value as an only input of the formulate can be in limited range $0.05 < \eta < 2$ for validating the formulate for numerical issues. The inverse Laplace transforms based on the classical de Hoog et al. algorithm will be used, as numerical inverse Laplace transform operation to convert the Laplace-space solution into a time-domain solution.

3.4.4. Fractional-in-time derivative equation ψ (t)

$$\tilde{\psi}(u) = \frac{1}{1 + u^{\beta}} \qquad 0 < \beta < 1 \quad (3.10)$$

Fokker-Planck Equation with Memory (FPME) described in previous sections can be reduced to fractional-in-time derivative equation (Berkowitz et al., 2002).

3.5. CTRW transport modeling for reactive contaminant

3.5.1. Introduction

Generally, transport of reactive solutes is modeled by ADE equation assuming Fickian transport condition with the long tailing of the BTCs just for adsorption effects. This effect of sorption can be brought to consideration in ADE formulate with use of retardation factor which defined by sorption kinetics or adsorption isotherms obtained from batch experiments, considering linear adsorption isotherm the retardation factor incorporates the distribution coefficient K_d.

$$\frac{\partial}{\partial x}(D_x \frac{\partial C}{\partial x}) - V_x \frac{\partial C}{\partial x} = R \frac{\partial C}{\partial t} \qquad (3.11)$$

Where D_x = hydrodynamic dispersion co-efficient, and v_x = velocity of contaminant through the porous medium and R = Retardation Coefficient.

Transport models based on ADE with a retardation coefficient which usually utilize the Fickian transport parameters v (fluid velocity) and D (tracer dispersion coefficient) often fail to interpret BTCs data and their long tails. Furthermore some previous studies have proven inaccuracy of parameters obtained from batch experiments may for application in transport models, because of different timescales in reaching equilibrium or the different degrees of solute mixing in the column or different mass ratios for chemical saturation (Maraqa et al., 1999; Nkedi- Kizza et al., 1989; Zhang and Selim, 2006).

For explaining the inadequacy of employing these parameters in models , different type of models have been developed based on making the link between reaction parameters and the flow rate under kinetic conditions (Pang et al., 2002, Zhang et al., 2008).These models have benefited the ADE with the addition of reaction by external retardation parameter, however one of the first assumption of these models is homogeneous porous medium that dictates Fickian transport and homogeneous mixing of solute, which often overestimates the overall reaction rate (Berkowitz et al., 2002).

It is proven that generally non-Fickian behavior and long tailing of BTCs in contaminant transport can be caused either by heterogeneities of the medium or by retardation of solutes due to adsorption/desorption processes or in case of suitable time scale both of them (Deng et al., 2008; Li and Ren, 2009). However the heterogeneity of the porous medium can affect also solute mixing procedure and sorption reaction (Edery et al., 2009, 2010). The first introduction of Continuous Time Random Walk (CTRW) method was by (Montroll and

Weiss, 1965). Before that the method was used to describe systems out of equilibrium, like anomalous diffusion phenomena which can be characterized by non– linear time dependence of mean squared displacement (MSD). The CTRW framework can account for a very wide range of non-Fickian and Fickian transport behaviours, with a smooth transition from non-Fickian to Fickian behaviour. As an very important fact, Berkowitz and Scher (1995,2001) for the first time introduced correlation between characteristic time scales (power-law transport time distributions) with distribution of heterogeneity length scales . Having the knowledge of previous CTRW modeling Dentz and Castro (2009) and Dentz and Bolster (2010) have proven the applicability of a random retardation field for CTRW.

3.5.2. Basic Formulation of Transport for reactive contaminant

CTRW general formulation can provide this possibility to model contaminant transport considering adsorption and desorption, by implementing some changes into the single transition pdf, $\tilde{\psi}$ (x,u) only. Proposed approach to treat sorption by Margolin et al. is based on applying a known function of a passive tracer, $\tilde{\psi}_0$, in the same domain. The new transition probability density function in Laplace domain is:

$$\tilde{\psi}(x, u) = \tilde{\psi}_0\big(x, u + \Lambda - \Lambda\tilde{\varphi}(u)\big) \quad (3.12)$$

Where $\tilde{\varphi}(u)$ is the Laplace transform of a single "sticking time" probability density function, Λ is the average **"sticking"** rate. In the Matlab CTRW ''toolbox'' developed by Cortis and Berkowitz (Cortis, A., and B. Berkowitz, 2005), the sticking time pdf was treated as a combination of two expressions: a user-defined function, and uniform behavior, weighted by W and (1-W) respectively, where W is provided by the user. For example, for a power law user-defined function:

$$\tilde{\varphi}(u) = W\frac{1}{1 + u^n} + (1 - W)\frac{1}{Tu} \quad (3.13)$$

Where n is a parameter of the user-defined function, and T represents the truncation time for the distribution.

3.6 USED CTRW TOOLBOX FEATURES

3.6.1 1-D Forward Modeling:

One dimensional Forward Modeling with introducing the parameters tabulated in table to the software, the CTRW model will be provided. Furthermore software has others options which allows the user to specify the choice of $\widetilde{\psi}(u)$ and the associated parameter values, the outlet and inlet boundary conditions, resident concentration (for a spatial distribution) or flux-averaged concentration (for a temporal breakthrough curve) solution, the (relative) length along the column at which the breakthrough curve is to be calculated and the parameters associated with the transport in the case of a reactive tracer.

3.6.2 1-D Inverse Problem: Fitting CTRW solutions to data sets

It is one of most useful feature of the software which mostly used in this study, consisting of a set of MATLAB m-files which can be used to obtain best fit parameters, given a set of concentration measurements from an experiment, for the CTRW models described in the previous sections. To solve the inverse problem, a convenient MATLAB structure that contains all relevant information on the experiments, the fitting parameters, and the fitted solutions was introduced. The fitting subroutine minimizes the norm of the difference between the data and the model. Several possible choices for the norm have been implemented, as well as the possibility of a logarithmic evaluation of the fits and the possibility to fit only parts of the data. All parameters are fit simultaneously. This can be a lengthy task, depending on how close the initial values of the parameters are to the optimized ones. Clearly, the better the initial guesses for the parameters, the faster the convergence of the minimization algorithm. Also, inaccurate estimation of the parameters may result in non convergence or convergence to a local minimum. For convenience, functions that provide graphical and textual output of the numerical results have been included in the toolbox.

CHAPTER 4

PHYSICAL AND NUMERICAL MODELING OF CONTAMINANT TRANSPORT

4.1 Physical modeling

The purpose of the present chapter on physical modelling is to study the scale-dependent dispersion in homogeneous soils for various observation scales, and also to study the effect of scale on type of soil. Therefore column tests simulating one-dimensional flow and dispersion were conducted in the laboratory on different types of soils based on details which are available in research work have been done by E. C. Nirmala Peter in 2007. The length dimension was chosen to allow a scale-dependent dispersion process to develop as much as possible. Four column tests were conducted. The details of experimentation, results and the break-through curves are presented in this chapter.

4.1.1 Materials used

a) Soil:

The details of experiments results which have done for determination of soil properties can be seen in table 4.1.

Table 4.1 Properties of soils used in column tests

S.No.	Type of soil	Gravel (%)	Coarse sand (%)	Medium sand (%)	Fine sand (%)	Silt (%)	Clay (%)	D_{50} mm	W L	PL	Specific Gravity (G)	Classification
1	CS_1	2.5	3.5	15	14	47.5	17.5	0.3	46	28	2.69	MI
2	CS_2	1	8	40	12	23	16	0.39	43	26	2.68	SC
3	CS_3	1	8	39	13	11	28	0.36	52	28	2.69	SC

b) Contaminant

Two types of contaminants, reactive and non-reactive, were used in the column tests. The non-reactive contaminant was a solution of Sodium Chloride with concentration approximately in the range of 3000 mg/L to 10000 mg/L. The reactive contaminant was a trivalent chromium solution with a concentration of 5000 mg/L. Two column tests were conducted using the non-reactive contaminant, and two tests with the reactive contaminant.

The concentration with which the contaminant was introduced into the column for each test is furnished in Table 4.2(b).

4.1.2   Experimental set up:

Two experimental set-ups A and B were used for conducting the column tests. Fig. 4.1 shows the set up A and B.

Fig.4.1 Experimental Set-Up A, 60 cm Long and 10 cm in Diameter four tests C_3 and C_4

(a)

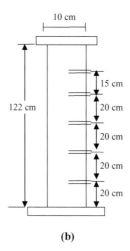

(b)

Fig.4.2 Experimental Set-Up B, (a) 122 cm long, and 10 cm in diameter mould (b) Schematic Diagram of the Mould with Perforated Tube Attached to the Inner Face of the Mould

4.1.3 Experimental program

Four column tests, C_1 through C_4 were planned keeping in view the objects of the project. Column tests C_1 and C_2 were conducted on comparatively long columns of lengths, 95 cm and 98 cm using set-up B Tests C_3 and C_4 were conducted using set-up A for reactive contaminant.

Table 4.2(a) Soil Type and Density conditions for the column tests

S.No.	Test No.	Soil type	Total length of column (cm)	Condition of soil column prepared			
				γ_b (t/m^3)	γ_d (t/m^3)	e	n (%)
1	C_1	CS_2	95	1.7	1.69	0.59	37
2	C_2	CS_3	98	1.7	1.63	0.67	40
3	C_3	CS_1	57	1.66	1.63	0.65	39
4	C_4	CS_1	57	1.474	1.45	0.83	45

Table 4.2 (b) Permeability and Seepage velocities of Soil column

S.No.	Test No.	Soil type	K (cm / s)	$v = k\,i$ (cm / s)	$v_x = v/n$ (cm / s)	Concentration of the contaminant C_0, mg/L
1	C_1	CS_2	5.16×10^{-3}	1.41×10^{-2}	3.81×10^{-2}	4527
2	C_2	CS_3	1.46×10^{-3}	3.87×10^{-3}	9.67×10^{-3}	3898
3	C_3	CS_1	3.07×10^{-5}	1.13×10^{-4}	2.90×10^{-4}	5000*
4	C_4	CS_1	4.95×10^{-5}	1.82×10^{-4}	4.02×10^{-4}	5000 *

*Chromium was used as contaminant where as for all the other experiments it was NaCl

4.1.4 Break-through curves

for tests C_1 and C_2, five to six sets of observations at different depths were recorded. The relative concentration, C/C_0 corresponding to each time interval at which the samples were collected, was calculated. The results are shown in Tables 4.3 through 4.7. A plot between time and relative concentration, i.e. break-through curve was drawn for each set of observations, and for each sample collection location. These break-through curves were utilized for the determination of the dispersion coefficients. The break-through curves are presented in chapter 6.

Table 4.3 Location of contaminant sample collection points

S.No.	Test	Total Length of the column (cm)	Depth of the sample collection point from soil surface (cm)
1	C1	95	15,35,55,75 and 95
2	C2	98	3,18,38,58,78 and 98
3	C3	57	57
4	C4	57	57

Table 4.4 Observed Relative concentrations with time for Test C_1
(γ_b=1.663g/cc, n = 38.4%, Chromium tracer)

S.No.	Time (min.)	C/C_0 At different distances (L) from the soil surface					
		0 cm	15 cm	35 cm	55 cm	75 cm	95 cm
1	0	1	0	0	0	0	0
2	15	1	0.0342	0.0243	0.023	0.015	0.012
3	30	1	0.189	0.161	0.143	0.123	0.104
4	45	1	0.397	0.331	0.287	0.243	0.221
5	60	1	0.508	0.453	0.419	0.353	0.32
6	75	1	0.585	0.53	0.464	0.431	0.386
7	90	1	0663	0.607	0.552	0.508	0.464
8	105	1	0.729	0.696	0.651	0.596	0.552
9	120	1	0.762	0.729	0.707	0.663	0.618
10	135	1	0.839	0.795	0.751	0.707	0.663
11	150	1	0.872	0.85	0.817	0.773	0.751
12	165	1	0.961	0.939	0.905	0.872	0.828
13	180	1	0.999	0.994	0.972	0.939	0.905

Table 4.5 Observed Relative concentrations with time for Test C_2

S.No.	Time (min.)	C/C_0 At different distances (L) from the soil surface					
		3 cm	18 cm	38 cm	58 cm	78 cm	98 cm
1	0	0	0	0	0	0	0
2	15	0.031	0.027	0.0245	0.023	0.0218	0.0192
3	30	0.203	0.175	0.15	0.11	0.109	0.102
4	45	0.421	0.368	0.326	0.286	0.27	0.232
5	60	0.49	0.42	0.38	0.35	0.34	0.326
6	75	0.704	0.623	0.53	0.488	0.421	0.407
7	90	0.825	0.69	0.596	0.501	0.46	0.434
8	105	0.865	0.744	0.609	0.515	0.488	0.461
9	120	0.895	0.892	0.731	0.663	0.528	0.488
10	135	0.933	0.93	0.825	0.677	0.542	0.528
11	150	1	1	0.852	0.798	0.69	0.623
12	165	1	1	0.92	0.858	0.744	0.717
13	180	1	1	1	1	0.865	0.798
14	195	1	1	1	1	1	0.879

Table 4.6 Observed Relative concentrations with time for Test C3
(γb=1.663g/cc, n = 38.4%, Chromium tracer)

S.No.	Time (hr)	C/C_0	S.No.	Time (min.)	C/C_0
1	1	0.0043	13	290	0.495
2	20	0.0186	14	310	0.615
3	26	0.023	15	334	0.669
4	45	0.024	16	358	0.717
5	50	0.028	17	378	0.738
6	69	0.029	18	402	0.807
7	117	0.0315	19	450	0.823
8	140	0.0329	20	474	0.865
9	165	0.078	21	498	0.872
10	188	0.1414	22	522	0.882
11	213	0.195	23	546	0.903
12	236	0.3595	24	620	0.956

Table 4.7 Observed Relative concentrations with time for Test C_4
(γ_b=1.474g/cc, n = 45.3%, Trivalent Chromium solution as contaminant)

S.No.	Time (hr)	C/C_0	S.No.	Time (hr)	C/C_0
1	0	0	9	192.5	0.516
2	1	0.024	10	220	0.625
3	23	0.029	11	240	0.70
4	24.25	0.033	12	260	0.775
5	70.17	0.104	13	280	0.84
6	118	0.281	14	290	0.875
7	142.3	0.355			
8	169	0.43			

4.2 Determination of hydrodynamic dispersion coefficient

Analytical models are commonly used in evaluating the transport parameters such as dispersion coefficient, retardation factor etc. from the laboratory column test results (Shackelford 1994, 1995, Shackelford and Redmond 1995).

In the classical Convection-Dispersion solution, the dispersivity is considered constant and it is a characteristic of a given medium. For One-Dimensional transport of linearly interacting solutes during steady-state water flow, the transport equation may be written as

$$R \, (\partial C / \partial t) = \partial / \partial x \, [D_x \, (\partial C / \partial x)] - v \, (\partial C / \partial x) - \mu \, C \qquad (4.1)$$

where C is the solute concentration, R is the retardation factor accounting for linear equilibrium sorption, D_x is the dispersion coefficient, v is the average steady state seepage velocity, μ is a first order decay coefficient, t is time, and x is distance. The dispersion coefficient D_x is generally considered to be a linear function of the pore water velocity as

$$D_x = D_d + \alpha \, v \qquad (4.2)$$

where D_d is the porous medium diffusion coefficient, and α is the dispersivity. For constant dispersivity, equation (4.1) reduces to the classical Convection-Dispersion Equation (CDE)

$$R \, (\partial C / \partial t) = D_x \, (\partial^2 C / \partial x^2)] - v \, (\partial C / \partial x) - \mu \, C \qquad (4.3)$$

The CDE model based on Eqs. (4.2) and (4.3) has been quite successful in describing results from laboratory displacement studies involving carefully constructed homogeneous soil columns. The dispersivity, α in such studies is usually of the order of a few millimeters or centimeters. These results are in contrast to those from field experimentation, which indicate that the dispersivity for transport in natural geologic media can be several orders of magnitude higher as compared with relatively small laboratory columns. Results from field studies also suggested that dispersivity or dispersion coefficient may be scale-dependent i.e. it increases with distance, x from the source of pollution. The growth with distance of the dispersion process may be a consequence of the heterogeneous nature of the sub surface environment. Barry and Sposito (1989) developed an analytical solution of the CDE with time-dependent transport coefficients. Yates (1990) developed an analytical solution for transport with scale-dependent dispersion by assuming that dispersivity increases linearly with distance. According to this model, increase in dispersivity with distance will be unlimited. Huang et al. (1995) developed a general analytical solution with scale-dependent dispersion. The solution assumes that the dispersivity, α increases linearly with distance, x that is $\alpha(x) = ax$ until some distance (xv), after which α reaches an asymptotic value, $\alpha_{L =}$

46

ax. Huang *et al.* (1995) conducted experimental investigation of solute transport in large homogeneous and heterogeneous saturated soil columns. The dispersion coefficient was calculated by fitting the analytical solution of Fractional Advection Dispersion Equation (FADE) developed by Zhou and Selim (2003) to the measured data at different transport scales. It was observed that the dispersion coefficient increased exponentially with transport scale for the homogeneous column, where as it increased with transport scale in a power law function for the heterogeneous column. The scale effect of the dispersion coefficient in the heterogeneous soil was much more significant compared to that in the homogeneous soil.

Wang *et al.* (1998) conducted column tests to obtain a better understanding of the relationship between hydrodynamic dispersion coefficient and seepage velocity. They used the analytical solution derived by Ogata and Banks (1961) and observed that if the calculated seepage velocity from Darcy's equation is used in the mathematical model, the analytical solution did not fit the experimental data well. This indicated that the true migration velocity might not be obtained by using the value of Darcy's velocity divided by porosity, n measured in the laboratory. From their experimentation it was also found that the use of measured seepage velocity, v_m in the mathematical model achieved a close fit to the experimental data. The measured seepage velocity is defined as the velocity of the mean point ($C/C_0 = 0.5$) of the Break-Through curve (Rumer, 1962). This finding has further confirmed the earlier works by Bigger and Nielson (1960). As contaminant migrates through the porous media, certain geochemical reactions occur between the c and the porous media. These interactions determine the relative rates at which chemicals migrate with respect to the advective front of water.

In the present work the classical Convection-Dispersion Equation (CDE) along with the measured mean seepage velocities (v_m) from the break-through curve for each sample collection location was used for the determination of dispersion coefficients. The measured mean seepage velocities for the reactive and non-reactive contaminants were used for the calculation of the retardation factors.

In the present study, the solution given by Fried (1975) was used to calculate dispersion coefficient. The classical Convection-Dispersion Equation considering only advection and dispersion of a non-reactive contaminant such as NaCl solution yields the following equation.

$$\partial C/\partial t = D \, (\partial^2 C/\partial x^2)] - v \, (\partial C/\partial x) \quad (4.4) \qquad \text{and}$$

$$R \, (\partial C/\partial t) = D \, (\partial^2 C/\partial x^2)] - v \, (\partial C/\partial x) \quad (4.5)$$

47

is the Convection-Dispersion with linear sorption for a reactive contaminant such as Trivalent Chromium solution.

A limited number of simple analytical solutions exist for one-dimensional problems with simplifying assumptions such as

1) The contaminant is ideal with constant density and viscosity
2) The fluid is incompressible
3) The medium is homogeneous and isotropic
4) Only saturated flow is considered.

Solutions to the CDE can be derived depending on initial and boundary conditions. In the present study, initial condition ($t = 0$), in the soil column the concentration of the contaminant was zero. For a continuous source of load, boundary conditions at the two ends of the one-dimensional soil column are,

$C (x, 0) = 0$

$C (0, t) = C_0$

$C (L, 0) = 0$

$C (L, t_\infty) = C_0$

The determination of hydrodynamic dispersion coefficient involves the following two steps.

a) Determination of measured seepage velocity (v_m)
b) Calculation of dispersion coefficient from analytical solution (D_x)
c) Determination of Retardation coefficient (R)

4.2.1 Determination of Measured (Mean) Velocities

Break-Through curves shown in figs 4.3 through 4.7 were used for the determination of the measured seepage velocities. From each curve, the time period corresponding to a relative concentration value of 0.5 ($C/C_0 = 0.5$) was obtained. The measured seepage velocity was determined as given below.

$$v_m = x / t_{0.5}$$

where   x = Distance in cm from the top surface of the column (source) to the point at which measurement of the concentration of the contaminant was taken.

$t_{0.5}$ = Time corresponding to a C/ C_0 of 0.5 from the Break-Through curve

4.2.2   Determination of Dispersion Coefficient from Analytical Solution

Fried (1975) suggested that in column tests measurements are very often performed at a given distance 'x' from the point of introduction of the contaminant, the longitudinal dispersion coefficient (D_x) is given by the formula,.

$$D_x = (1/8) \left\{ \left[(x - v\, t_{0.16})/ (t_{0.16})^{0.5} \right] - \left[(x - v\, t_{0.84})/ (t_{0.84})^{0.5} \right] \right\}^{2} \qquad (4.6)$$

v = Darcy's seepage velocity

Wang et al (1998) suggested from their experimentation that, the use of measured seepage velocity (v_m) in the mathematical model achieved a close fit to the experimental data. Therefore in the present work, measured seepage velocities (v_m) were used instead of Darcy's seepage velocities (v) in equation (4.6). For tests C_3, C_4 the measurement of concentration of the contaminant was performed at only one location that is the bottom of the column, where as for tests C_1, C_2 the contaminant samples were collected at different locations along the length of the column, Table 4.8. The dispersion coefficient values for C_1, C_2 tests indicate variation of the coefficient values with the distance, x. This further emphasizes that dispersion coefficient is a scale-dependent value. Measurements of time $t_{0.16}$ and $t_{0.84}$ were also made corresponding to relative concentrations of 0.16 and 0.84 respectively.

Table 4.8 Measured Velocities and Dispersion Coefficients of the Soils

S.No.	Test	*Distance (cm)	$t_{0.16}$ (s)	$t_{0.50}$ (s)	$t_{0.84}$ (s)	V_x (cm/s)	D_x (cm^2/s)
1	C_1	15	1475.28	4303	7499.34	0.0035	0.0186
		35	1844.1	4671.72	7745.22	0.0075	0.0713
		55	1967.04	4917.6	8482.86	0.0112	0.1734
		75	2212.92	75778.2	9097.56	0.013	0.2579
		95	2458.8	5778.2	9589.32	0.0164	0.3776
2	C_2	3	1598.936	3289.45	5526.28	0.0009	0.00055
		18	1776.303	3947.34	6578.9	0.00456	0.0183
		38	1973.67	4407.86	8552.57	0.0086	0.092
		58	2368.4	5263.12	8947.30	0.011	0.147
		78	2434.2	6315.74	9868.35	0.0123	0.249
		98	2499.98	6710.48	10526.24	0.0146	0.393
3	C_3	57	630000	1044000	1638000	0.000054	0.000358
4	C_4	57	330000	690000	1010000	0.000083	0.000765

*Distance of the contaminant sample collection location from soil top surface i.e. source

4.3 Spatial variation of dispersion coefficient:

As Power law variation of dispersion coefficient with length were found the most effective and accurate mode for modeling the spatial variation of dispersion coefficient by E. C. Nirmala Peter in 2007 , in this study this mode was used in numerical modelling.

4.3.1 Power Law Variation

For power law variation of dispersion coefficient, D_x, with 'x', the following expression was used.

$$D_x = D_d + m\, x^n$$

where m and n are the parameters dependent on the type of the soil.

$$D_x - D_d = m\, x^n$$

Taking logarithms on both sides

$$\text{Log } (D_x - D_d) = \text{Log } (m\ x^n)$$

$$\text{Log } (D_x - D_d) = \log m + n \text{ Log } x \quad (4.7)$$

For tests C_1, C_2, the plots were drawn for Log $(Dx - Dd)$ versus Log x (Figs. 4.4 through 4.6). The diffusion coefficient (D_d) was assumed as 1.5×10^{-6} cm^2/s (based on free solution diffusion coefficients after Shackelford 1989, Shackelford and Daniel, 1991). The intercept of the line obtained from the plot gives the Log m value, and the slope of the line gives the n value. The values of 'm' and 'n' obtained are 0.0002 and 1.6466; 0.00075 and 1.8827; and 0.00009 and 1.8618 for C_1, C_2. Table 4.3 furnishes the values of 'm' and 'n'.

Table 4.9 Parameters "m and n" for Power Law Variation of Dispersion Coefficient with length

S.No.	Test	M	n
1	C_1	0.00020	1.6466
2	C_2	0.00075	1.8750

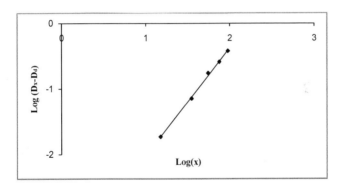

4.3 Power Law Variation of D_x with Length x for C_1

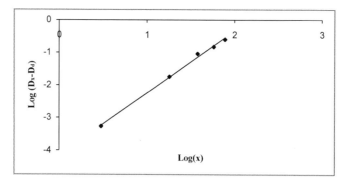

4.4 Power Law Variation of D_x with Length x for C_2

4.4 Numerical modeling of one dimensional contaminant transport

In the present work numerical model was developed considering power law variations of dispersion coefficient with distance 'x' from the contaminant source for both non-reactive and reactive contaminants. Based on the study which has been done by E.C.NIRMALA PETER, 2007 the model has been developed with brief following equations and assumptions, comprehensive discussion can be seen in elsewhere.

a)   Convective-Advective Equation (CDE) for non-reactive contaminant

$$\frac{\partial}{\partial x}(D_x \frac{\partial C}{\partial x}) - V_x \frac{\partial C}{\partial x} = \frac{\partial C}{\partial t} \qquad (4.8)$$

and for a reactive contaminant $\frac{\partial}{\partial x}(D_x \frac{\partial C}{\partial x}) - V_x \frac{\partial C}{\partial x} = R \frac{\partial C}{\partial t}$ (4.9)

Where D_x = hydrodynamic dispersion co-efficient, and v_x = velocity of contaminant through the porous medium and R = Retardation Coefficient.

b)   The initial and boundary conditions are:

C (x, 0) = 0 at t = 0

C (0, t) = C_0 for t > 0

C (L_1, t) = C (L, t) L_1>L

where L is the total depth or thickness of the stratum.

c) Convective-Dispersive Equation (CDE) for Scale-Dependent Dispersion

$$\left(\frac{D_d}{LV_x} + \frac{m.X^n L^{n-1}}{V_x}\right)\frac{\partial^2 C}{\partial X^2} - \left(\frac{m.n.X^{n-1}L^{n-1}}{V_x} - 1\right)\frac{\partial C}{\partial X} = \frac{\partial C}{\partial T} \tag{4.10}$$

d) Convective-Dispersive Equation with Sorption for Scale-Dependent Dispersion

$$\left(\frac{D_d}{R.L^2 V_x} + \frac{m.X^n L^{n-1}}{RV_x}\right)\frac{\partial^2 C}{\partial X^2} + \left(\frac{m.n.X^n .L^{n-1}}{R.V_x} - \frac{1}{R}\right)\frac{\partial C}{\partial X} = \frac{\partial C}{\partial T} \tag{4.11}$$

e) Finite Difference Method of Solution to the Governing Equation

$$C_{i,t+\Delta t} = \left[\frac{(\alpha_D + m.X)}{(\Delta x)^2}.\Delta T + \frac{(m-1)}{2.\Delta x}.\Delta T\right].C_{i-1,t} + \left(1 - 2.\frac{(\alpha_D + m.X)}{(\Delta x)^2}.\Delta T\right).C_{i,t} +$$

$$\left[\frac{(\alpha_D + m.X).\Delta T}{(\Delta x)^2} - \frac{(m-1)}{2.\Delta x}.\Delta T\right].C_{i+1,t} \tag{4.12}$$

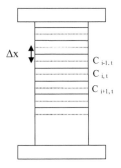

Fig. 4.5 Elements for Finite Difference Analysis

CHAPTER 5

SAMPLING-BASED SENSITIVITY ANALYSIS FOR MODELS

5.1 Introduction

Sensitivity analysis (SA) is the study of how the uncertainty in the output of a model (numerical or otherwise) can be apportioned to different sources of uncertainty in the model input. This study can clarify the robustness of the model predictions and also simplify models.

It can also Support decision making or the development of recommendations for decision makers; Enhancing communication from modelers to decision makers; Increased understanding or quantification of the system ; and Model development .It can also provide very useful information about Factors that mostly contribute to the output variability.

The region in the space of input factors for which the model output is either maximum or minimum or within pre-defined bounds, Optimal or instability regions within the space of factors for use in a subsequent calibration study, Interaction between factors.

One of the most prevalent used methodologies of sensitivity analyses is sampling-based which can be implemented without access to model equations or even the model code .The model is executed repeatedly for combinations of values sampled from the distribution (assumed known) of the input factors, and a relationship between inputs and outputs is established using the model results at the sample points. In the absence of information on joint probabilities, a common simplifying assumption is that the parameters are independent (Saltelli et al. (2000)).

5.2 Sensitivity analysis for CTRW truncated power law method

As it is discussed in chapter 3, equation (3.6) truncated power law methods requires 3 input parameters β, t_1, t_2, where $\tau_2 = t_2/t_1$ (where t_1 and t_2 have dimensions of time). The time t_1 represents the approximate median transition time and sets the lower limit from which the power law behavior begins. The presence of a cut-off time t_2 ensures that for $t > t_2$ the transport evolves into a normal one. Hence anomalous transport in any system resides in the interplay between the duration of the observation and the extent of disorder, which is reflected in the range of t up to the cut-off t_2. The exponent β is a very useful means to characterize the latter range. Here with respect to these parameters as our inputs parameters and breakthrough curves which have been obtained from our ADE and CTRW modeling as output parameters, the sensitivity analysis have been done in two ways:

5.2.1 Effect of β on BTC for different values of τ

CTRW modeling CTRW modeling is carried out with velocity v= 0.1 cm/s and dispersion coefficient D=0.005 cm2/s as input parameters with the boundary condition of C (x, 0) = 0 at t = 0, C (0, t) = C0 for t > 0, C (L1, t) = C (L, t) L_1>L, which is discussed in chapter 4 and for different values of β=0.25, 0.5, 0.75, 1.25, 1.5, 1.75, 2.25, 2.5, 2.75 and τ=10, 100, and 1000 to study the effect of β on the output break through curves. The study was graphically presented in Fig.5.1 through 5.8.

In Figs. 5.1 – 5.3, the break-through curve (BTC) using ADE in the figure indicates fickian transport. The curves for β = 0.25, 0.5 and 0.75 for different values of τ (10,100 and 1000) are presented and $\tau = t_2/t_1$ where t_1 denotes a typical median transition time and t_2 corresponds to the largest heterogeneity length scale. In order to study the quantitative changes in the character of the special concentration profiles in the transition from anomalous to normal (fickian) behavior, the values of τ are varied. The effect of v and D provide valuable information about the shape of the BTC and special shape of BTC.

In fig. 5.1 for β = 0.25, the BTCs deviated from ADE more and more with increase in τ value. However with increase in β (0.5 and 0.75), the deviations of BTCs decreased and the time required to achieve a relative concentration of almost 1 also decreased (Figs. 5.2 & 5.3). Therefore it is observed that for β<1, as τ increases the anomalous nature of transport becomes more predominant. The effect of τ decreased with increase in the values of β (1.25, 1.5 and 1.75; Figs. 5.4 – 5.6) for $1 < \beta < 2$. The break-through curves moved closer to ADE and the deviation of BTCs for different τ values also decreased and the curves moved closer and became sharper. The time scale required for achieving the required relative concentration also decreased with increase in β value. For β =1.75, the BTCs for τ = 10, 100 and 1000 almost merged with ADE. Figs 5.7-5.9 shows that for β > 2, the curves for τ = 100 and 1000 merged indicating the effect of heterogeneity length scale ceases and the transport behavior appears to be Fickian and normal. It is also observed that for β > 2, the time scale to reach a relative concentration of 1 remained almost same all the curves.

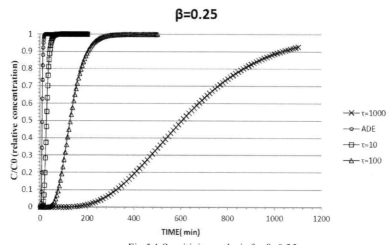

Fig.5.1 Sensitivity analysis for β=0.25

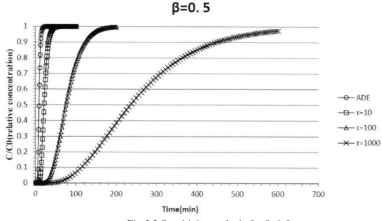

Fig.5.2 Sensitivity analysis for β=0.5

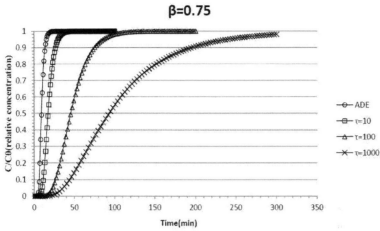

Fig.5.3 Sensitivity analysis for β=0.75

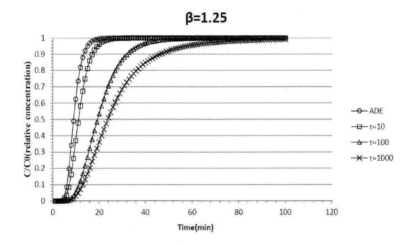

Fig.5.4 Sensitivity analysis for β=1.25

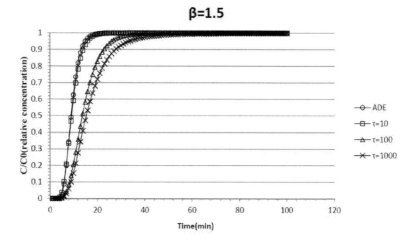

Fig.5.5 Sensitivity analysis for β=1.5

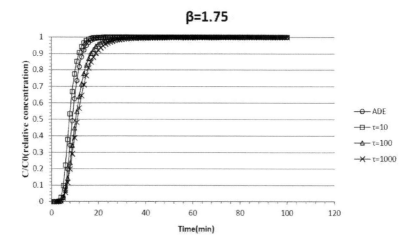

Fig.5.6 Sensitivity analysis for β=1.75

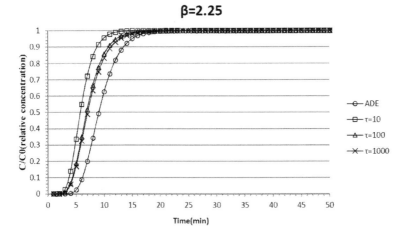

Fig.5.7 Sensitivity analysis for β=2.25

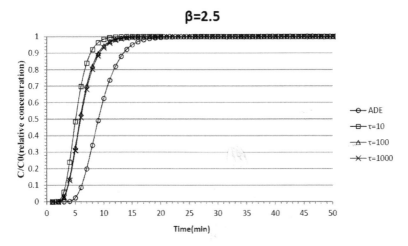

Fig.5.8 Sensitivity analysis for β=2.5

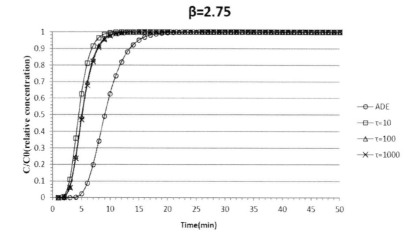

Fig.5.9 Sensitivity analysis for β=2.75

5.2.2 Effect of velocity on the BTC

For studying the effect of change in velocity and type of flow in CTRW modeling, sensitivity analysis was done with respect to the experimental data obtained from physical modeling, the column test results with velocity v= 0.1, 0.03, 0.004, 0.0004 (cm/s) and dispersion coefficient D=0.005, 0.012, 0.0003, 0.0001 cm^2/s with boundary condition of C $(x, 0) = 0$ at $t = 0$, C $(0, t) = C_0$ for $t > 0$, C $(L_1, t) = C (L, t)$ $L_1 > L$, which is discussed in chapter 4 was considered and the results of changing the input parameters (β=0.5, 1.5, 2.5) for τ=100 in out puts data break through curves are studied and graphically presented in Fig.5.10 through 5.13. In another case for the constant β value break through curves were drawn for different V and D values which are presented in Fig.5.14 through 5.16.

Figs. 5.10 through 5.13 shows, for a given velocity and τ value (i. e. 100), the time scale increases with decrease in β value. With decrease in the velocity, the time scale for a required relative concentration increases. It is also observed that for $\beta < 1$, late arrivals and long tails were obtained while very steep curves were obtained for $\beta > 1$. The steepness of the curves increased with increase in β value. In these figures irrespective of the values of v and D, the BTCs for $\beta = 0.5$ are showing considerable deviation from the ADE and the curves for β between 1.5 and 2.5 lies on either side of ADE. Especially for $\beta = 2.5$ and with low velocity, the BTC is very close to ADE. Lower values of β indicate more heterogeneity. The pdf increases with decrease in the value of β.

The same behavior can be observed in last three figures about increasing β value for different V and D values but it is remarkable that,as our expectation, decreasing the velocity and dispersion coefficient causes more dispersive behavior of curves, and breakthrough times become longer in such a way that for increasing the V value from 0.1cm/s to 0.0004cm/s the longest breakthrough time which is for β=0.5 has changed from 200 min to 30000min.

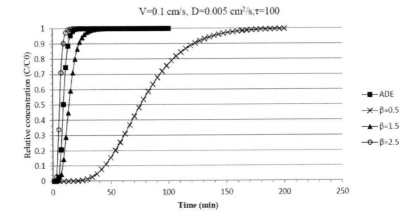

Fig.5.10 variable β value sensitivity analysis for
V=0.1cm/s, D=0.005cm^2/s,τ=100

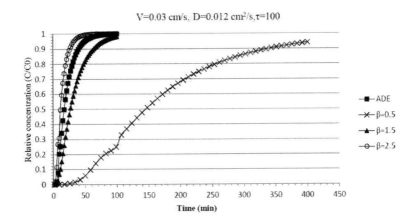

Fig.5.11 variable β value sensitivity analysis for
V=0.03cm/s, D=0.012cm^2/s,τ=100

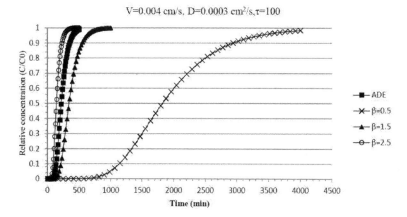

Fig.5.12 variable β value sensitivity analysis for
V=0.004cm/s, D=0.0003cm^2/s,τ=100

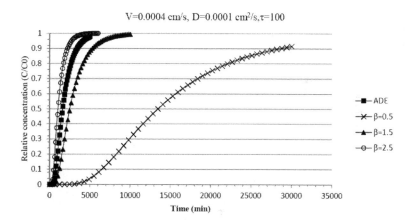

Fig.5.13 variable β value sensitivity analysis for
V=0.0004cm/s, D=0.0001cm^2/s,τ=100

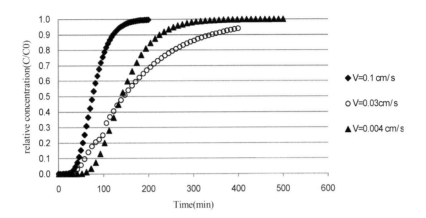

Fig.5.14 variable velocity and dispersion sensitivity analysis for
$\beta=0.5$,$\tau=100$

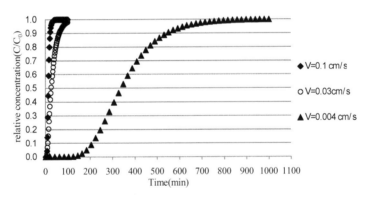

Fig.5.15 variable velocity and dispersion sensitivity analysis for
$\beta=1.5$,$\tau=100$

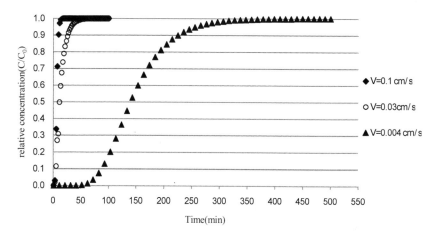

Fig.5.16 variable velocity and dispersion sensitivity analysis for
$\beta=2.5$,$\tau=100$

5.3 Conclusion

Sensitivity analysis for the CTRW model was carried out by studying the effect of the parameters β, τ, v and D.

1. The effect of changing in $\tau_2 = t_2/t_1$ is most predominant for $\beta <1$, as t_2 corresponding to the heterogeneous length scale increases the anomalous or non-fickian nature of transport also increases.

2. For $\beta < 1$, smaller the value of β more will be the time scale required to achieve a relative concentration of 1.

3. As β increases from 1 to 2, the effect of variation of τ decreases and the BTCs also become sharper move close to ADE.

4. As the effect of τ ceases for $\beta > 2$, the transport behavior changes from anomalous to fickian condition

5. Irrespective of the velocity and dispersion coefficient values, the BTCs show considerable deviation from ADE for $\beta < 0.5$. The curves also indicate late arrivals and long tails.

6. With decrease in the velocity, the time scale increases irrespective of the β values.

7. All the parameters remaining the same, if Dispersion coefficient (D) increases, the time scale decreases considerably for $\beta < 1$, slight variation for $1 < \beta < 2$ and almost no variation for $\beta > 2$.

CHAPTER 6

VERIFICATION AND COMPARISON OF THE MODELS

6.1) Comparison the usage of seepage velocity with mean velocity in CTRW models

As it is discussed in chapter 4 Wang et al. (1998) conducted column tests to obtain a better understanding of the relationship between hydrodynamic dispersion coefficient and seepage velocity. They used the analytical solution derived by Ogata and Banks (1961) and observed that if the calculated seepage velocity from Darcy's equation is used in the mathematical model, the analytical solution did not fit the experimental data well. This indicated that the true migration velocity might not be obtained by using the value of Darcy's velocity divided by porosity, n measured in the laboratory. From their experimentation it was also found that the use of measured seepage velocity, V_m in the mathematical model achieved a close fit to the experimental data. The measured seepage velocity is defined as the velocity of the mean point ($C/C_0 = 0.5$) of the Break-Through curve (Rumer, 1962). This finding has further verified the earlier works by Bigger and Nielson (1960) and E. C. Nirmala Peter (2007). In the present work CTRW method was applied once with the measured mean seepage velocities (v_m) calculated from the break-through curve and once for seepage velocity for each sample collection location in tests C_1 and C_2 to compare the results of each models. 1-D Forward Modelling feature of software was used to develop the CTRW models and draw the BTCs, for determination of suitable inputs parameters vast trial error procedure was done by author and the results were compared by experimental breakthrough curves ,and the best one presented in following tables and figures.

Table 6.1 Measured Velocities of test C_1

TEST C_1			
Location No	Distance(cm)*	$v_x = v/n$ (cm/s)	$v_m = x / t_{0.5}$(cm/s)
1	15	0.0381	0.0035
2	35	0.0381	0.0075
3	55	0.0381	0.0112
4	75	0.0381	0.013
5	95	0.0381	0.0164

*Distance of the contaminant sample collection location from soil top surface i.e. source

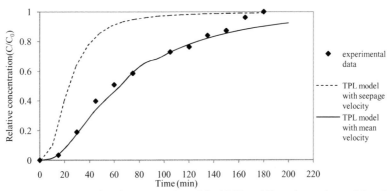

Fig 6.1 BTC comparison between TPL model with V_x and V_m and experimental data for test C_1 in 15 cm depth

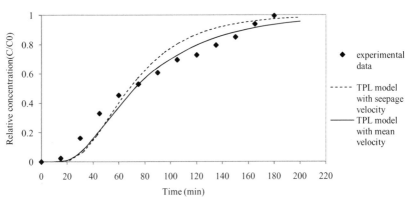

Fig 6.2 BTC comparison between TPL model with V_x and V_m and experimental data for test C_1 in 35 cm depth

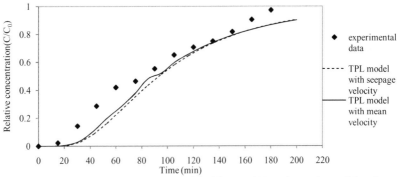

Fig 6.3 BTC comparison between TPL model with V_x and V_m and experimental data for test C_1 in 55 cm depth

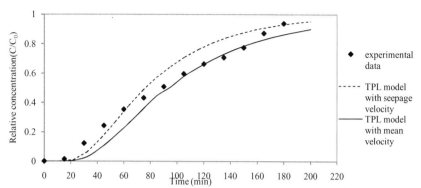

Fig 6.4 BTC comparison between TPL model with V_x and V_m and experimental data for test C_1 in 75 cm depth

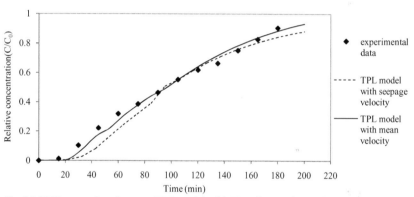

Fig 6.5 BTC comparison between TPL model with V_x and V_m and experimental data for test C_1 in 95 cm depth

Table 6.2 Measured Velocities of test C_2

TEST C_2			
Location No	Distance(cm)	$v_x = v/n$ (cm/s)	$v_m = x / t_{0.5}$ (cm/s)
1	3	9.67×10^{-3}	0.0009
2	18	9.67×10^{-3}	0.00456
3	38	9.67×10^{-3}	0.0086
4	58	9.67×10^{-3}	0.011
5	78	9.67×10^{-3}	0.0123
6	98	9.67×10^{-3}	0.0146

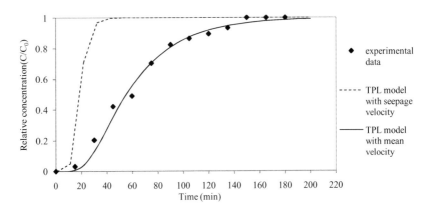

Fig 6.6 BTC comparison between TPL model with V_x and V_m and experimental data for test C_2 in 3 cm depth

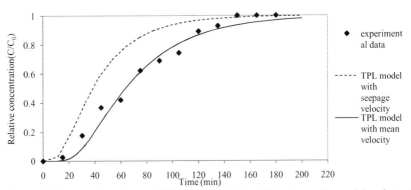

Fig 6.7 BTC comparison between TPL model with V_x and V_m and experimental data for test C_2 in 18 cm depth

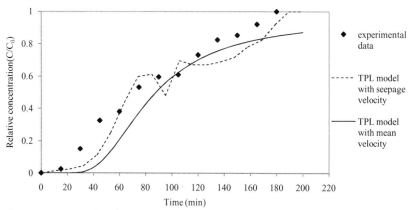

Fig 6.8 BTC comparison between TPL model with V_x and V_m and experimental data for test C_2 in 38 cm depth

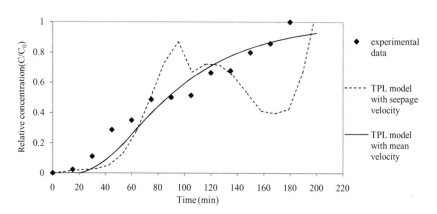

Fig 6.9 BTC comparison between TPL model with V_x and V_m and experimental data for test C_2 in 58 cm depth

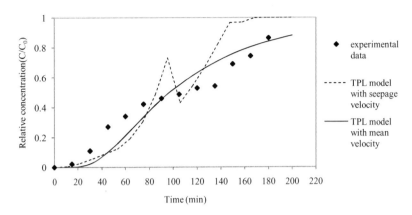

Fig 6.10 BTC comparison between TPL model with V_x and V_m and experimental data for test C_2 in 78 cm depth

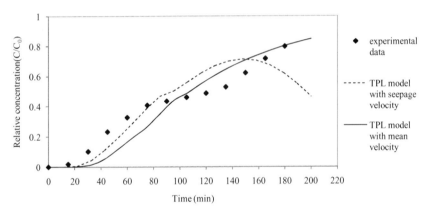

Fig 6.11 BTC comparison between TPL model with V_x and V_m and experimental data for test C_2 in 98 cm depth

➢ In the entire figure it is completely clear that TPL models with use of mean velocity can predict the experimental results much better than same model using the seepage velocity. In figures 6.8 to 6.10 kinks and irregularity can be observed in TPL models with seepage velocity mainly because numerical error in solution of partial differential equation of CTRW caused by introducing extraneous data with respect to dispersion coefficient. Based on the introduced results and graphs, it is concluded that using mean velocity instead of average seepage velocity can remarkably improve the modelling results, therefore for all CTRW and ADE modelling, mean velocity was used in this study.

6.2) Comparison dispersion coefficient obtained from CTRW and ADE models and experimental values

As it is mentioned in chapter 4 for tests C_1 and C_2 dispersion coefficients were calculated by the formula developed by Fried (1975) equation (4.6) with the assumption of Wang et al (1998) to use mean velocity instead of seepage velocity and the results are tabulated in table 4.1 .

In CTRW models as it is discussed in section 4, there is possibility to use very useful feature of toolbox for solving the inverse problems and Fitting CTRW solutions to data sets and determined the best fitting parameters .

In this section the comparison between dispersion coefficients obtained from analytical models and CTRW truncated power law method and ADE method with use of mean velocity has been done and results are tabulated in table 6.3 and 6.5 and represented graphically in figures 6.12 and 6.15. For the better comparison between coefficient obtained from different methods, power law variation parameters were obtained and compared with the same parameters which have been used in finite difference models. The results can be seen in fig.6.13, 6.14 and 6.16, 6.17 and table 6.4 and 6.6.

Table 6.3 Measured Dispersion Coefficients of test C_1

TEST C_1				
Location No	Distance(cm)	Dx (cm^2/s)	$D_{ADE}(cm^2/s)$	$D_{TPL}(cm^2/s)$
1	15	0.0186	0.0135	0.0247
2	35	0.0713	0.0673	0.0857
3	55	0.1734	0.1512	0.2268
4	75	0.2579	0.2531	0.3187
5	95	0.3776	0.3760	0.5715

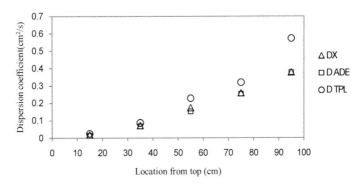

Fig 6.12 Dispersion coefficient obtained from different method versus depth for C_1

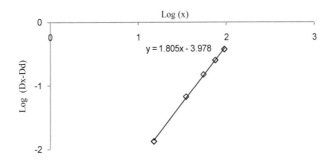

Fig 6.13 Power Law Variation of D_{ADE} with Length x for C_1

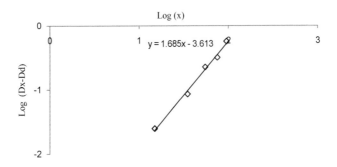

Fig 6.14 Power Law Variation of D_{TPL} with Length x for C_1

Table 6.4 Parameters "m and n" for Power Law Variation of Dispersion Coefficient with length for all methods for C_1

Modeling Methods	M	n
Analytical	0.00020	1.6466
ADE	0.000105	1.805
TPL	0.000244	1.685

Table 6.5 Measured Dispersion Coefficients of C_2

TEST C2				
Location No	Distance(cm)	D_x (cm^2/s)	D_{ADE}(cm^2/s)	D_{TPL}(cm^2/s)
1	3	0.00055	0.00066	0.000765
2	18	0.0183	0.02052	0.0216
3	38	0.092	0.09145	0.09738
4	58	0.149	0.14577	0.15138
5	78	0.249	0.23322	0.26364
6	98	0.393	0.35214	0.44818

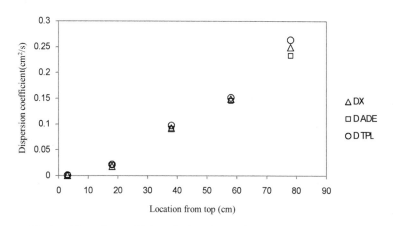

Fig 6.15 Dispersion coefficient obtained from different method versus depth for C_2

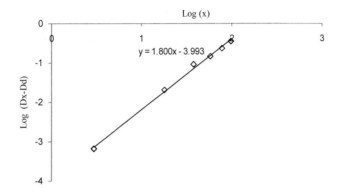

Fig.6.16 Power Law Variation of D_{ADE} with Length x for C_2

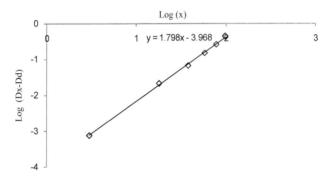

Fig.6.17 Power Law Variation of D_{TPL} with Length x for C_2

Table 6.6 Parameters "m and n" for Power Law Variation of
Dispersion Coefficient with length for test C_2

Modeling Methods	M	n
Analytical	0.00075	1.8750
ADE	0.0001015	1.8006
TPL	0.00010762	1.685

➤ The figures 6.12 and 6.15 indicate that dispersion coefficients predicted by TPL and ADE and analytical (Fried (1975)) methods are almost equal in shallow depth of sampling (below 30 cm), while the values will be diverged in more depth in such a way that TPL methods introduce greater value than ADE and analytical methods. Consequently the dispersion coefficient can be simply obtained by analytical solution are valid just for the movement of contaminant in shallow depth of soil. The figures 6.13, 6.14, 6.16 and 6.17 indicate the change in the parameters of power law variation which were used in finite difference modeling.

6.3) Verification of finite difference models with variable dispersion coefficient and comparison all the methods for non reactive contaminant

Verification of the designed model involves testing the model's ability to reproduce another set of measurements using the model parameters that were developed in the calibration process. However in the present study the model was verified by using the parameters from the laboratory tests. The dispersion coefficients calculated from the laboratory column tests using analytical solution indicated variation in dispersion coefficient with distance of travel along the length of the column. Therefore in the present study with respect to the results obtained from power law variations of the dispersion coefficient was considered. The verification of the designed model was carried out by determining a set of model input parameters from the data obtained through the analytical solution. These parameters were used in the numerical models developed for power law variations of dispersion coefficient to check or verify whether the measured concentrations for the column can be approximated. The Break-Through curves obtained from the above procedure were also compared with the curves obtained from ADE and CTRW model in next section. In the present study the verification of CDM for power law variation of dispersion coefficient for a non-reactive contaminant was carried out by developing break-through curves (Numerical curves) through numerical modelling using parameters such as m, n and V_x obtained from the column experiments for tests C_1, C_2. These numerical curves were compared with the respective break-through curves from the laboratory observations and other models. Verification of CDM for power law variation of dispersion coefficient was also conducted for the published experimental data of Wang et al. (1998) for short columns and Huang et al. (1995) for long homogeneous columns. The verification of the linear variation model was conducted by developing numerical curves from the parameters m and V_x from the experimental.

6.3.1) Input parameters for the development of numerical modelling for power law variation of D_x with distance and verification models for tests C_1 and C_2

The input parameters Vx, αD, β, m and n for power law variation of the dispersion coefficient are taken from laboratory tests (Tables 4.1 and 4.4). Input parameters for power law variable dispersion coefficient finite difference model for the full-length of the columns of 95 cm in C_1, 98 cm in C_2 are furnished in Tables 6.7 through 6.8 .The break-through curves are given combined by other methods break through curves in Figures 6.18 through 6.28.

Table 6.7 Input Parameters for Power Law Variation of Dispersion Coefficient with Length of the Column for Test C_1

Distance of measurement (cm)	15	35	55	75	95
Dd (cm2/s)	1.5x10-6	1.5x10-6	1.5x10-6	1.5x10-6	1.5x10-6
M	0.0002	0.0002	0.0002	0.0002	0.0002
N	1.6466	1.6466	1.6466	1.6466	1.6466
Vx(cm/s)	0.0035	0.0075	0.0112	0.013	0.0164
α D	2.86x10-6	5.71x10-6	2.44x10-6	1.54x10-6	9.63x10-7
β	0.329171	0.26568	0.238304	0.250892	0.231707

Table 6.8 Input Parameters for Power Law Variation of Dispersion Coefficient with Length of the Column for Test C_2

Distance of measurement (cm)	3	18	38	58	78	98
Dd(cm2/s)	1.5x10-6	1.5x10-6	1.5x10-6	1.5x10-6	1.5x10-6	1.5x10-6
m	0.000079	0.000079	0.000079	0.000079	0.000079	0.000079
n	1.8750	1.8750	1.8750	1.8750	1.8750	1.8750
Vx(cm/s)	0.0009	4.56E-03	8.60E-03	0.011	0.0123	0.0146
α D	5.56x10-4	1.83x10-5	4.59x10-6	2.35x10-6	1.56x10-6	1.05x10-6
β	0.22954	0.21728	0.22153	0.25071	0.29060	0.298955

6.3.2) Input parameters for the development of CTRW truncated power law model for tests C_1 and C_2

It was mentioned in chapter 3 and specifically in section 3.6, to obtain solutions in CTRW method in software toolbox for forward problems, the user should input specific parameter values, the inlet boundary condition, and the choice of $\tilde{\psi}(t)$ (i.e., the Laplace transform of $\psi(t)$, and the toolbox calculates either a 1D temporal or 1D/2D spatial concentration profile of the migrating solute. Therefore in this study for test C_1 and C_2, truncated power law function were chosen as Pdf and based on the results came in this chapter in section 6.1 for each test mean velocity V_m and analytical dispersion coefficient D_x were inputted as specific parameters and other parameters were obtained by vast trial and error procedure which have done by author which tabulated in tables 6.9 and 6.10.

Table 6.9 Input Parameters for CTRW truncated Power Law method for Test C_1

Test C_1				
Location No	Distance(cm)	β value	t_1(min)	t_2(min)
1	15	1.6622	0.512861	1.262E+11
2	35	1.6595	0.003613	2793186.9
3	55	1.6539	0.009945	19801587
4	75	1.6491	0.005254	16737863
5	95	1.6421	0.004631	3684681.9

Table 6.10 Input Parameters for CTRW truncated Power Law method for Test C_2

Test C_2				
Location No	Distance(cm)	β value	t_1(min)	t_2(min)
1	3	1.6876	0.003595837	13070734.71
2	18	1.6874	0.006160274	29295438.3
3	38	1.6737	0.003136173	232380670.3
4	58	1.6633	0.000774819	16413454.67
5	78	1.6746	0.0626037	40681.78498
6	98	1.6404	0.000772325	23062161.02

6.3.3) Comparison the results for C_1 and C_2

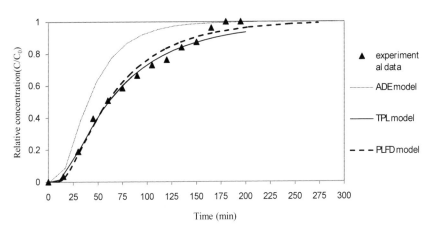

Fig.6.18 BTCs for comparison between power law finite difference, CTRW (TPL) and ADE model, with experimental data for test C_1 in 15 cm depth

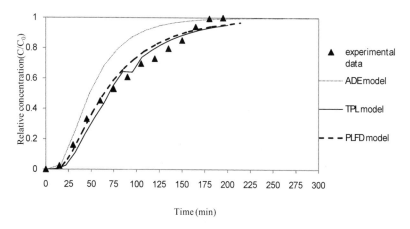

Fig.6.19 BTCs for comparison between power law finite difference, CTRW (TPL) and ADE model, with experimental data for test C_1 in 35 cm depth

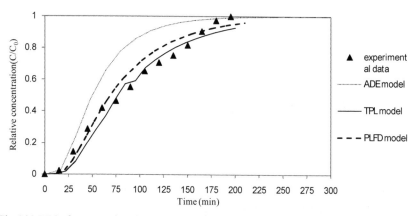

Fig.6.20 BTCs for comparison between power law finite difference, CTRW (TPL) and ADE model, with experimental data for test C_1 in 55 cm depth

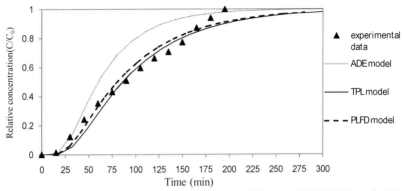

Fig.6.21 BTCs for comparison between power law finite difference, CTRW (TPL) and ADE model, with experimental data for test C_1 in 75 cm depth

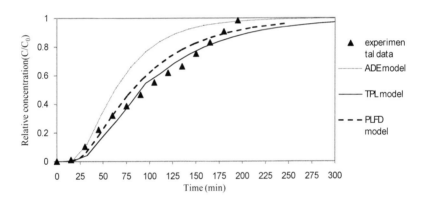

Fig.6.22 BTCs for comparison between power law finite difference, CTRW (TPL) and ADE model, with experimental data for test C_1 in 95 cm depth

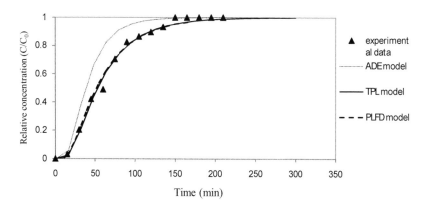

Fig.6.23 BTCs for comparison between power law finite difference, CTRW (TPL) and ADE model, with experimental data for test C_2 in 3 cm depth

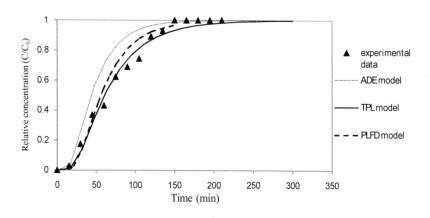

Fig.6.24 BTCs for comparison between power law finite difference, CTRW (TPL) and ADE model, with experimental data for test C_2 in 18 cm depth

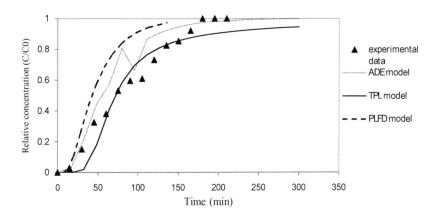

Fig.6.25 BTCs for comparison between power law finite difference, CTRW (TPL) and ADE model, with experimental data for test C_2 in 38 cm depth

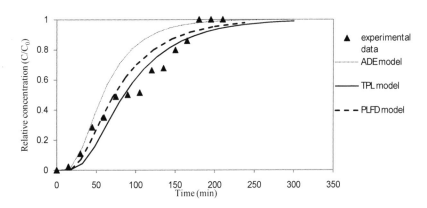

Fig.6.26 BTCs for comparison between power law finite difference, CTRW (TPL) and ADE model, with experimental data for test C_2 in 58 cm depth

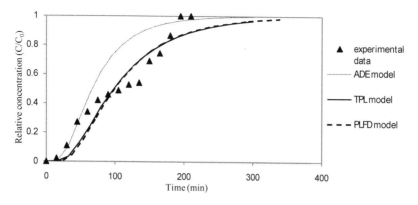

Fig.6.27 BTCs for comparison between power law finite difference, CTRW (TPL) and ADE model, with experimental data for test C_2 in 78 cm depth

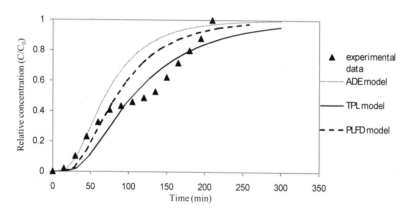

Fig.6.28 BTCs for comparison between power law finite difference, CTRW (TPL) and ADE model, with experimental data for test C_2 in 98 cm depth

➢ All the figures indicate that ADE solutions over-predict the concentrations, and the finite difference model with spatial power law variation of dispersion coefficient predicted the concentrations at most of the locations satisfactorily, while truncated power law CTRW model measurements, were in the best agreement with the experimentally measured concentration data in all the cases. Figure 6.23 and 6.27 show that in test C_2 for the depth of 3cm and 78cm TPL breakthrough curve is converged to the PLFD curve which can be for similarities of the power law finite difference formulate and the truncated power law formulate in those depths. In figures 6.25 kink and irregularity can be observed in ADE mainly because numerical error in solution ADE equation in CTRW toolbox.

6.3.4 Verification of finite difference model and input parameters for numerical model using experimental data of Huang *et al.* (1995)

Laboratory tracer (NaCl) experiments were conducted to investigate solute transport in 12.5 m long, horizontally placed homogeneous sand column and break-through curves were obtained.

The concentration of the contaminant in the column was measured with electrical conductivity probes inserted at every 50 cm interval. Fig. 6.29 shows the observed and analytically fitted concentration distributions at several locations in the column for tracer injection in the homogeneous column (Huang *et al.* 1995).

The relative concentrations of the contaminant for various time periods at each of the locations (2 m, 5 m, 8 m and 11 m) were taken out from the Fig.6.29. This data of times and relative concentrations is furnished in Table 6.11. Verification of CDM for power law variation of dispersion coefficient was carried out by developing break-through curves at different locations of the column, i.e. 2 m, 5 m, 8 m and 11 m.

The seepage velocities calculated from the ratio of the distance of travel (2 m, 5 m, 8 m and 11 m) to the time corresponding to a relative concentration of 0.5 ($t_{0.5}$) obtained from the BTC's of the laboratory data were used for predicting the numerical curves.

The input parameters for the numerical solutions are furnished in Table 6.12. The 'm' and 'n' values obtained from log-log plot of D_x (Fried, 1975) and distance, x, are 5.01×10^{-6} and 1.5946 respectively. However it was observed that the numerical solutions using these parameters deviated from the laboratory concentration profiles, which may be due to the inaccuracies involved in the measurement of the relative concentrations from the Fig. 6.29 (published). So the model was calibrated to obtain solutions with values of 'm' equal to

5.01×10^{-6}, 2.9×10^{-6}, 2.15×10^{-6} and 2×10^{-6}, and an 'n' value of 1.5635 for 2 m, 5 m, 8 m, and 11 m lengths of the column respectively, which fitted well with the experimental data.

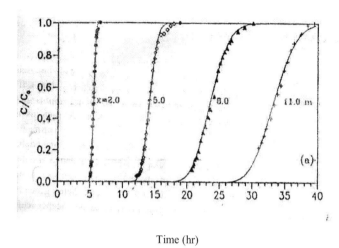

Time (hr)

Fig. 6.29 Observed (symbols) concentration Distribution at Several Locations in the Homogeneous Column (Huang et al. 1995)

Table 6.11 Experimental Data for Different Locations of the Homogeneous Column from Huang *et al.* (1995)

x = 2 m		x = 5 m		x = 8 m		x = 11 m	
Time(hr)	C/C_0	Time(hr)	C/C_0	Time(hr)	C/C_0	Time(hr)	C/C_0
4.81	0	12.282	0.0317	20.747	0.071	30.871	0.246
4.979	0.0159	12.614	0.0476	20.913	0.126	31.701	0.3094
5.145	0.0476	12.946	0.0763	22.075	0.253	32.697	0.4285
5.228	0.0952	13.278	0.1428	22.987	0.404	33.527	05634
5.311	0.2142	13.444	0.222	24.398	06348	34.522	0.7062
5.643	0.349	13.61	0.3015	25.228	0.785	35.601	0.8173
5.71	0.5633	13.942	0.3967	26.556	0.896	37.012	0.9364
5.726	0.738	14.274	0.492	28.548	0.983	39.004	0.999
5.783	0.8094	14.439	0.6268	30.041	0.999		
5.809	0.912	14.77	0.7538				
6.307	0.9681	15.269	0.881				
6.781	0.9998	15.601	0.9284				
		17.759	0998				

Table 6.12 Input Parameters for the Development of Break-Through Curves at Different Locations of the Homogeneous Column

Depth, x	2 m	5 m	8 m	11 m
D_d (cm^2/s)	1.5E-06	1.5E-06	1.5E-06	1.5E-06
m	4.85E-06	2.9E-06	2.15E-06	2E-06
n	1.5635	1.5635	1.5635	1.5635
X	1	1	1	1
X^n	1	1	1	1
$X^{(n-1)}$	1	1	1	1
L (cm)	200	500	800	1100
$L^{(n-1)}$	19.798	33.18	43.24	51.74
V_x	0.009921	0.0098	0.009416	0.0092
ΔX	0.0196	0.0196	0.0196	0.0196
ΔT	0.01	0.015	0.015	0.015
α_D	7.56E-06	3.06E-07	1.99E-07	1.48E-07
β	0.009678	0.009819	0.009873	0.01125
μ_x	0.25196	0.3834	0.38552	0.4392
β_x	-0.25124	-0.37678	-0.37675	-0.3759
$\mu_x + \beta_x$	0.000717	0.006611	0.00877	0.06327
$1 - 2\mu_x$	0.4961	0.2332	0.22896	0.12162
$\mu_x - \beta_x$	0.5032	0.7602	0.76226	0.81511

89

6.3.5) Input parameters for the development of CTRW truncated power law model for using experimental data of Huang *et al*. (1995)

For CTRW modelling using experimental data of Huang et al. (1995), truncated power law function were chosen as Pdf and based on the results came in this chapter in section 8.1 for each test mean velocity V_m and analytical dispersion coefficient D_x were inputted as specific parameters and other parameters were obtained by vast trial and error procedure which have done by author which tabulated in table 8.13.

Table 6.13 Input Parameters for CTRW truncated Power Law method for experimental data of Huang et al. (1995)

Huang *et al*. (1995)				
Location No	Distance(m)	β value	t_1(min)	t_2(min)
1	2	2.5992	0.002068	9.1791E+14
2	5	2.196	0.00302	7735707
3	8	2.042	0.047665	65042.92
4	11	1.9994	0.00266	2.28034E+14

6.3.6) Comparison the results for experimental data of Huang et al. (1995):

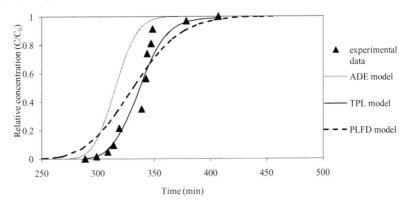

Fig.6.30 BTCs for comparison between power law finite difference, CTRW (TPL) and ADE model, with experimental data for Huang test in 2 m depth

90

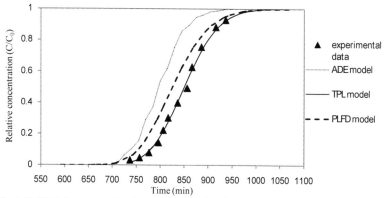

Fig.6.31 BTCs for comparison between power law finite difference, CTRW (TPL) and ADE
model, with experimental data for Huang test in 5 m depth

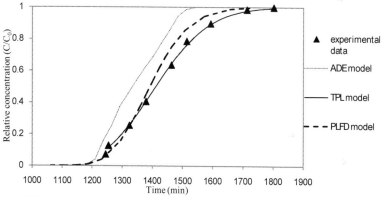

Fig.6.32 BTCs for comparison between power law finite difference, CTRW (TPL) and ADE
model, with experimental data for Huang test in 8 m depth

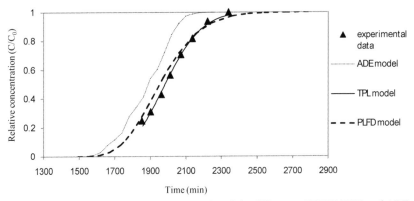

Fig.6.33 BTCs for comparison between power law finite difference, CTRW (TPL) and ADE
model, with experimental data for Huang test in 11 m depth

➢ All the figures indicate that ADE solutions over-predict the concentrations, and the finite
difference model with spatial power law variation of dispersion coefficient predicted the
concentrations at most of the locations satisfactorily, while truncated power law CTRW
model measurements, were in the best agreement with the experimentally measured
concentration data in all the cases. With comparison of the figures it can be concluded
that with increasing the depth of sampling from 3m to 11m the break through curves
become flatter and more disperse, consequently breakthrough time will be increased.

6.3.7 Verification of finite difference model and input parameters for numerical model using experimental data of WANG et al. (1998)

Wang et al. (1998) conducted a series of experiments on short sand columns
(L = 46.5 cm) to study the relationship between the dispersion coefficient and seepage
velocity. Two types of column tests (experiments A and B) were conducted. The first type of
column tests (experiment A) was used to determine the value of dispersion coefficient of the
medium and the second set of column tests (experiment B) was performed to verify the value
of dispersion coefficient obtained from the first set of column tests. In experiments A, a
conductivity probe was used to measure the concentration of the contaminant (NaCl solution)
at 20 cm below the top of the sand sample. Tests were conducted at different flow rates and
each column test A_1, A_{10}, A_4 and A_5 correspond to a specific discharge of 0.6159, 0.3663,
0.2319 and 0.2036 cm³/s respectively. Break-through curves were developed using the

analytic solution derived by Ogata and Banks (1961) for this one-dimensional migration of the contaminant in the column test. The relative concentrations of the contaminant at various time periods for each of these columns A_1, A_{10}, A_4 and A_5 were taken from the Fig.8.34 and this data of time versus relative concentration is furnished in Table 8.14. Verification of CDM for power law variation of dispersion coefficient was carried out by developing solutions to the respective columns A_1, A_{10}, A_4 and A_5 taking length, L equal to 20 cm and seepage velocity V_x equal to the ratio of length (L) to time corresponding to a relative concentration of 0.5 ($t_{0.5}$) from the BTCs for the laboratory test data. The input parameters for the numerical solutions are furnished in Table 6.15.

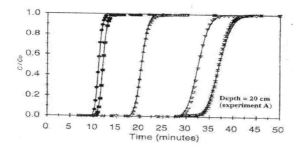

Fig. 6.34 Comparison of Numerical Solutions with Experimental Data of Wang et al. (1998)

Table 6.14 Experimental Data for A_1, A_{10}, A_4 and A_5 from Wang *et al.* (1998)

Expt. A_1		Expt. A_{10}		Expt. A_4		Expt. A_5	
Time, min	C/C_0	Time, min	C/C_0	Time, min	C/C_0	Time, min	C/C_0
10.545	0	17.76	0	28.86	0	32.19	0
10.822	0.078	17.87	0.0377	29.97	0.0472	33.577	0.0472
11.10	0.085	19.147	0.0849	30.94	0.0755	34.41	0.1038
11.655	0.160	19.425	0.1226	31.08	0.1415	35.104	0.1698
11.932	4	19.564	0.1887	31.22	0.1887	35.797	0.2925
12.21	0.292	19.702	0.283	31.357	0.2359	36.53	0.4623
12.487	4	19.98	0.3679	31.635	0.2831	37.185	0.632
12.765	0.443	20.257	0.443	31.912	0.3491	38.29	0.7548
13.04	4	20.535	0.5378	32.467	0.472	38.85	0.868
13.597	0.575	20.8125	0.623	32.745	0.585	39.96	0.934
14.43	5	21.09	0.679	33.3	0.7076	42.18	0.981
	0.707	21.23	0.755	33.855	0.783		
	0.877	21.645	0.811	34.41	0.849		
	0.924	21.922	0.858	34.6875	0.9058		
	6	22.20	0.896	34.955	0.9435		
	0.981	22.755	0.934	35.797	0.9529		
		23.865	0.981	38.296	0.981		

Table 6.15 Input Parameters for the Development of Break- Through Curves   for A_1, A_{10}, A_4 and A_5

Experiment	A_1	A_{10}	A_4	A_5
D_d (cm^2/s)	1.5×10^{-6}	1.5×10^{-6}	1.5×10^{-6}	1.5×10^{-6}
m	2.45×10^{-5}	8.65×10^{-6}	8.99×10^{-6}	4.5×10^{-6}
n	1.8	1.975	1.8	1.9975
L	20	20	20	20
V_x(cm/s)	0.0267	0.0166	0.01026	0.00889
α_D	2.81×10^{-6}	4.52×10^{-6}	7.31×10^{-6}	8.44×10^{-6}
β	0.010075	0.00966	0.009625	0.010048

6.3.8) Input parameters for the development of CTRW truncated power law model for using experimental data of WANG *et al.* (1998)

For CTRW modelling using experimental data of WANG *et al.* (1998), truncated power law function were chosen as Pdf and based on the results came in this chapter in section 6.1 for each test mean velocity V_m and analytical dispersion coefficient D_x were inputted as specific parameters and other parameters were obtained by vast trial and error procedure which have done by author which tabulated in table 6.16.

Table 6.16 Input Parameters for CTRW truncated Power Law method for experimental data of Huang et al. (1995)

WANG *et al* test			
Test ID	β value	t_1(min)	t_2(min)
A_1	1.8952	1.58708E-06	1.72624E+15
A_{10}	1.785	1.44977E-05	6.28348E+17
A_4	1.6829	1.40929E-05	7.92866E+11
A_5	1.6321	4.70111E-06	2.432204

6.3.9) Comparison the results for experimental data of WANG *et al.* (1998)

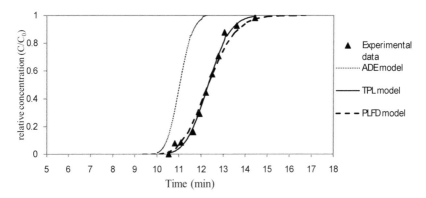

Fig.6.35 BTCs for comparison between power law finite difference, CTRW (TPL) and ADE model, with experimental data for WANG et al, A_1

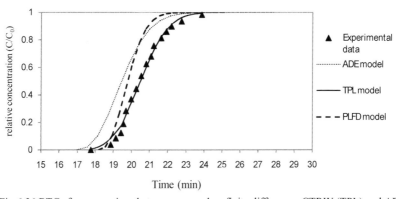

Fig.6.36 BTCs for comparison between power law finite difference, CTRW (TPL) and ADE model, with experimental data for WANG et al, A_{10}

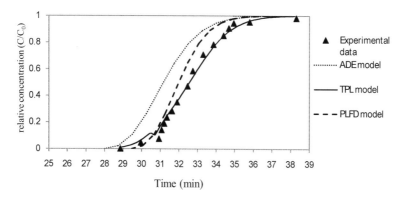

Fig.6.37 BTCs for comparison between power law finite difference, CTRW (TPL) and ADE model, with experimental data for WANG et al,A_4

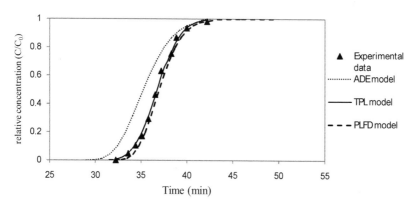

Fig.6.38 BTCs for comparison between power law finite difference, CTRW (TPL) and ADE model, with experimental data for WANG et al, A_5

➤ All the figures indicate that ADE solutions over-predict the concentrations, and the finite difference model with spatial power law variation of dispersion coefficient predicted the concentrations at most of the locations satisfactorily, while truncated power law CTRW model measurements, were in the best agreement with the experimentally measured concentration data in all the cases.

6.4 VERIFICATION OF THE MODELS FOR CONVECTIVE- DISPERSIVE TRANSPORT WITH SORPTION

The tests C_3 and C_4 were conducted on the soil CS_2 with only a single contaminant collection point provided below the bottom of the soil columns. Therefore values of 'm' and 'n' for power law variation could not be obtained from the laboratory tests. However the break-through curves are drawn assuming these parameters from the calibration carried out earlier. Tables 6.17 and 6.18 show the input parameters for the curves. Break-through curves are given in Figures 6.22 and 6.23.

Table 6.17 Input Parameters for CTRW truncated Power Law method for

CTRW input parameters			
Test ID	sorption parameters $[\Lambda, T, W]$	sorbing function in $\tilde{\varphi}(u)$	parameters of $\tilde{\varphi}(u)$ $[\beta, t_1, t_2]$
C_3	[0.2, 100, 1]	TPL	[1.8997, 0.00058, 21973538.5]
C_4	[4.7, 100, 1]	TPL	[1.6468, 0.000072, 7556140070.9]

Table 6.18 Input Parameters for Power Law Variation of Dispersion
Coefficient with Length of the Column Test C_3 and C_4

Parameters	C_3	C_4
D_x (cm^2/ s)	0.000358	0.000765
D_d (cm^2/ s)	0.000003	0.000003
L (cm)	57	57
V_x (cm/ s)	0.000054	0.000083
m	0.00005	0.00003
n	1.357	1.5
R	6.39	4.57
$L^{(n-1)}$	4.235	7.55
α_D	0.000975	0.000634
β	3.921296	2.728916
α_R	0.156495	0.218818
X	1	1
X^n	1	1
ΔX	0.0196	0.0196
ΔT	0.00025	0.00025
μ_x	0.399452	0.388689
β_x	0.004313	0.004317
$\mu_x + \beta_x$	0.403765	0.393006
$1 - 2\mu_x$	0.201096	0.222621
$\mu_x - \beta_x$	0.395139	0.384373

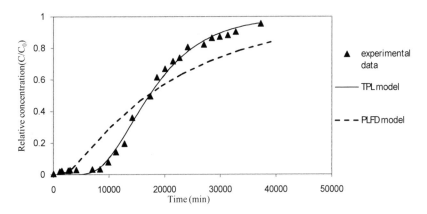

Fig.6.39 BTCs for comparison between power law finite difference, CTRW (TPL) model, with experimental data for reactive contaminant C_3

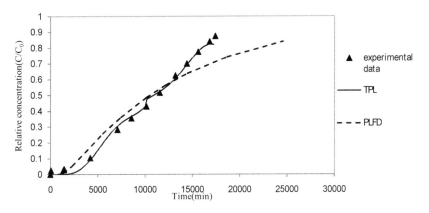

Fig.6.40 BTCs for comparison between power law finite difference, CTRW (TPL) model, with experimental data for reactive contaminant C_4

➤ Two figures indicate that the developed finite difference models with power law variations of dispersion coefficient with distance could not be verified in full-fledged manner due to lack of experimental or published data providing the contaminant concentration profiles at different lengths of the column, while truncated power law CTRW model measurements, were in the best agreement with the experimentally measured concentration data in all the cases.

6.5 Conclusion

From the comparison between using seepage velocity and mean velocity in CTRW methods it can be concluded that TPL models with use of mean velocity can predict the experimental results much better than same model using the seepage velocity. In some of the curves kinks and irregularity can be observed in TPL models with seepage velocity mainly because numerical error in solution of partial differential equation of CTRW caused by introducing extraneous data with respect to dispersion coefficient.

Based on the second part of this study ,dispersion coefficients predicted by TPL and ADE and analytical (Fried (1975)) methods are almost equal in shallow depth of sampling (below 30 cm), while the values will be diverged in more depth in such a way that TPL methods introduce greater value than ADE and analytical methods. Consequently the dispersion coefficient can be simply obtained by analytical solution are valid just for the movement of contaminant in shallow depth of soil.

Based on the comparison between models, ADE solutions over-predict the concentrations, and the finite difference model with spatial power law variation of dispersion coefficient predicted the concentrations at most of the locations satisfactorily, while truncated power law CTRW model measurements, were in the best agreement with the experimentally measured concentration data in all the cases.

With comparison of the figures it can be concluded that with increasing the depth of sampling, the break through curves become flatter and more disperse, consequently breakthrough time will be increased.

CHAPTER 7

CONCLUSIONS

Considering a wide variety of modeling approaches that attempt to predict the spatial and temporal migration of chemical species either conservative or reactive in geological formations this research has been motivated by finding the best and the most accurate method for modeling the contaminant transport. Most of previous approaches are based on various deterministic and stochastic forms of the so-called advection-dispersion equation (ADE) with the basic and crucial assumption that dispersion (plume spreading) behaves macroscopically as a Fickian diffusive process, with the macro dispersivity being assumed constant in space and time. However, laboratory, field, and stochastic analyses have demonstrated that dispersivity as a variable term which is indeed dependent on the time and/or length scale of measurement. This fact usually leads the research about different approaches of contaminant transport modeling to anomalous transport term.

In the present study for better understanding of anomalous transport phenomena, based on physical modeling which was carried out in the laboratory on two sand columns (C_1, C_2) of moderate lengths of 94 to 98 cm with non-reactive contaminants and in two columns(C_3, C_4) of length 57 cm with reactive contaminants, the dispersion coefficients were obtained initially from the break-through curves using the classical Fickian type of CDE.

Furthermore, measured seepage velocity (v_m) defined as the velocity of the mean point (C/C_0 = 0.5) of the Break-Through curve was used instead of Darcy's seepage velocity (v_x) for the determination of the dispersion coefficient from Fride's (1975) solution. It was established from these results that the dispersion coefficient is a function of distance of travel. The models have been developed and solved using the finite difference method for the one-dimensional flow and dispersion considering the spatial variations based on power law ($D_x = D_0 + mx^n$), Numerical experimentation (parametric study) was carried out by varying the different relevant parameters within the practical ranges.

On the other hand the truncated power law CTRW modelling were developed by use of Matlab CTRW "toolbox" developed by Cortis and Berkowitz. The solutions obtained from the proposed models were verified and compared with the data observed in the laboratory column tests, and also for the published experimental data of Wang *et al.* (1998) and long column experimental data of Huang *et al.* (1995). Sampling based sensitivity analysis for finite difference and CTRW methods were carried out; the comparison between CTRW models with use of seepage velocity and mean velocity was also done to evaluate accuracy of Ogata and Banks (1961) assumption.

The comparison between dispersion coefficients obtained from analytical models and CTRW truncated power law method and ADE method with use of mean velocity has been done, and finally for all the four tests in addition to experimental data obtained from Huang et al. (1995) and WANG et al. (1998) verification of CTRW and Finite difference models and comparison of Break-through curves corresponding to CTRW, ADE and finite difference methods with experimental data have been done in this study. Based on the results of this study following points can be concluded:

1. For all the tests with non reactive contaminants, that ADE solutions over-predict the concentrations, and the finite difference model with spatial power law variation of dispersion coefficient predicted the concentrations at most of the locations satisfactorily, while truncated power law CTRW model measurements, were in the best agreement with the experimentally measured concentration data in all the cases.

2. For reactive contaminant, it was also observed that the numerical solution is not in good agreement with the measured concentration data, and the reason may be attributed to nonlinear adsorption of the contaminant due to finite adsorption site which can be counted as the deficiency of the present numerical method, while in truncated power law CTRW model, very good agreement was observed between predicted model and experimental data.

3. According to the Comparison of the usage of seepage velocity with mean velocity in CTRW models, it can be concluded that the use of measured seepage velocity (v_m) obtained from break-through curves with Ogata and Banks (1961) method, in the CTRW model achieved very much close fit to the experimental data compare to the same models using seepage velocity.

4. With use of MATLAB structure that contains all relevant information on the experiments, the fitting parameters, and the fitted solutions, dispersion coefficients were calculated for all collection points for tests C_1 and C_2. The results indicate that dispersion coefficients predicted by CTRW and ADE and analytical (Fried (1975)) methods are almost equal in shallow depth of sampling (below 30 cm), while the values will be diverged in more depth in such a way that CTRW methods introduce greater value than ADE and analytical methods. Consequently

the dispersion coefficient obtained simply by analytical solution can be valid just for the movement of contaminant in the shallow depth of soil.

5. For the first time the quantitative coefficient were implemented for determination of nature of contaminant transport in laboratory column tests. According to literature reviews and sensitivity analysis which have been done by author, in the case of using truncated power law method in CTRW the most important and effective parameter is the exponent β. For β >2, the CTRW model yields the classical Fickian behaviour described by the ADE model. For $1 <\beta< 2$, the mean of the tracer plume moves with a constant velocity. The curves are asymmetric with long late time tails and as increases the resulting BTCs become sharper and less disperse. For $0 < \beta <1$, the BTCs display the most anomalous behaviour. The curves are not symmetrical and delayed early and long late time tails exist. The shapes of the BTCs are functions of β, and are similar on different spatial scales.

6. For tests C_1, C_2, C_3, C_4 and short column experimental data of Wang in all the sampling locations the condition was $1 <\beta< 2$ which means, mean of the tracer plume moves with a constant velocity. The curves are asymmetric with long late time tails and as increases the resulting BTCs become sharper and less disperse, While in the case of long column introduced in Haung experimental data for the sampling location before 11 m the condition β > 2 the CTRW model yields the classical Fickian behaviour described by the ADE model and Gaussian distribution is recovered, but for the location11 m the $1<\beta<2$, which means for sufficiently long travel distances, the transport nature changes gradually from fickian to non fickian.

7. The full transport behavior for a truncated power-law transition time distribution, which is characterized by a power-law behavior in an intermediate time regime defined by the time scales t_1 and t_2 have been investigated. The time scale t_1 characterizes a typical transition time on a local length scale, while t_2 corresponds to the largest heterogeneity length scale. For transition times large compared to t_2 the (t) decreases exponentially. Within the presented conceptual model, solute transport times which are larger than t_2 correspond to largest heterogeneity length scale. Thus, for $t \gg t_2$ the medium appears homogeneous and the solute transport is normal. For transport times which are small compared to t_1, the solute has not yet sampled the smallest heterogeneity scales and the CTRW approach has only a formal

meaning. In experimental study as t_2 is larger than the experiment duration. When the cutoff time t_2 is large the mass transfer rate required to capture the tailing behavior is low.

8. Based on the results obtained from modeling of tests C_1, C_2, C_3, C_4 and short column experimental data of Wang and long column of Haung experimental data clearly it can be concluded that with increasing the depth of sampling the β value gradually decrease.

9. Based on the sensitivity analysis results, it can be concluded that the effect of τ is most predominant for $\beta < 1$. As t_2 corresponding to the heterogeneous length scale increases the anomalous or non-Fickian nature of transport also increases. For $\beta < 1$, smaller the value of β more will be the time scale required to achieve a relative concentration of 1. As β increases from 1 to 2, the effect of variation of τ decreases and the BTCs also become sharper move close to ADE. As the effect of τ ceases for $\beta > 2$, the transport behavior changes from anomalous to Fickian condition, Irrespective of the velocity and dispersion coefficient values, the BTCs show considerable deviation from ADE for $\beta < 0.5$. The curves also indicate late arrivals and long tails. With decrease in the velocity, the time scale increases irrespective of the β values. All the parameters remaining same, if Dispersion coefficient (D) increases, the time scale decreases considerably for $\beta < 1$, slight variation for $1 < \beta < 2$ and almost no variation for $\beta > 2$.

CHAPTER 8

FUTURE SCOPS

1) The CTRW framework offers a viable and general means to model tracer transport, which is more accurate than descriptions obtained with the conventional ADE and other approaches. This general framework can guide the design of new experiments on tracer displacement. Systematic experimental studies and analyses on the effects of pore structure, fluid velocity, column length, and water saturation on BTCs will, in particular, help to shed additional light on the relationship between characteristics and specific parameters of the CTRW theory.

2) Development of physical and experimental two dimensional solute transport models and study the reliability of CTRW toolbox for two dimensional contaminant movements.

3) Continuing the comparison study with use of other stochastic methods like stochastic solute transport model (SSTM) developed by Kulasiri and Verwoerd.

4) Incorporating more accurate methods like finite element method for the same assumptions in one dimensional solute transport or use of software's like MODFLOW, Pullute-V7, Vleach-V2.2, 2Dfatmic-V1, etc and comparison the results with present finite difference model and CTRW.

5) Use of soft computing methods like neural network for prediction the CTRW parameters based on the vast calibration data collected from future experiments, and improving the present Matlab toolbox.

6) In case of reactive contaminant more comprehensive experimental study with higher concentration of introduced solute should be done to improve physical modeling in the case of dealing with sorption.

REFERENCES

1. Abramowitz, M., and I. Stegun (1970), Handbook of Mathematical Functions, Dover, Mineola, N. Y.

2. Adams, E.E., and L.W. Gelhar. 1992. Field study of dispersion in a heterogeneous aquifer, 2. Spatial moment analysis. Water Resour.Res. 28:3293–3308.

3. Aronofsky, J. S., and J. P. Heller (1957), A diffusion model to explain mixing of flowing miscible fluids in porous media, Trans. Am. Inst. Min. Metall. Pet. Eng., 210, 345–349.

4. Bear, J. 1972. Dynamics of fluids in porous media. American Elsevier, New York. drol. 47:29–51.

5. Berkowitz, B., A. Cortis, M. Dentz, and H. Scher (2006), Modeling non-Fickian transport in geological formations as a continuous time random walk, Rev. Geophys., 44, RG2003.

6. Berkowitz, B., G. Kosakowski, G. Margolin, and H. Scher. 2001. Application of continuous time random walk theory to tracer test measurements in fractured and heterogeneous porous media. Ground Water ,39:593–604

7. Berkowitz, B., and H. Scher (1995), On characterization of anomalous dispersion in porous and fractured media, Water Resour. Res., 31(6), 1461– 1466.

8. B. Berkowitz, J. Klafter, R. Metzler, H. Scher, Physical pictures of transport in heterogeneous media: advection–dispersion, random walk and fractional derivative formulations, Water Resource. Res. 38 (10) (2002) 1191.

9. Berkowitz, B., and H. Scher. 1998. Theory of anomalous chemical transport in random fracture networks. Phys. Rev. E 57:5858–5869.

10. Berkowitz, B.,H. Scher, and S.E. Silliman. 2000. Anomalous transport in laboratory-scale, heterogeneous porous media. Water Resour.Res. 36:149–158.

11. Berkowitz, B., and H. Scher. 2001. The role of probabilistic approaches to transport theory in heterogeneous media. Transport Porous Med. 42:241–263.

12. Bos, F. C., and D. M. Burland (1987), Hole transport in poly vinyl carbazole— The vital importance of excitation-light intensity, Phys. Rev. Lett., 58(2), 152–155.

13. Boggs, J.M., S.C. Young, and L.C. Beard. 1992. Field study of dispersion in a heterogeneous aquifer. 1. Overview and site description. Water Resour. Res. 28:3281–3291.

14. Burr, D.T., and E.A. Sudicky. 1994. Nonreactive and reactive solute transport in three-dimensional heterogeneous porous media: Mean displacement, plume spreading, and uncertainty. Water Resour. Res. 30:791–815.

15. Carrera, J., X. Sa´nchez-Vila, I. Benet, A. Medina, G. Galarza, and J. Guimera` . 1998. On matrix diffusion: Formulations, solution methods, and qualitative effects. Hydrogeol. J. 6:178–190.

16. Cherny, I.A., 1963. Groundwater hydrology. Gostoptechizdat, page 396.

17. Cortis, A., and B. Berkowitz (2004), Anomalous transport in "classical" soil and sand columns, Soil Sci. Soc. Am. J., 68, 1539– 1548.

18. Cortis, A., and B. Berkowitz (2005), Computing 'anomalous' contaminant transport in porous media: The CTRW MATLAB toolbox, Ground Water, 43(6), 947–950.

19. Cortis, A., C. Gallo, H. Scher, and B. Berkowitz (2004b), Numerical simulation of non-Fickian transport in geological formations with multiple-scale heterogeneities, Water Resour. Res., 40.

20. Cortis, A., Y. Chen, H. Scher, and B. Berkowitz (2004a), Quantitative characterization of pore-scale disorder effects on transport in "homogeneous" granular media, Phys. Rev. E, 70(10),

21. Cunningham, J.A., C.J. Werth, M. Reinhard, and P.V. Roberts. 1997. Effects of grain-scale mass transfer on the transport of volatile organics through sediments: 1. Model development. Water Resour. Res. 33:2713–2726.

22. Dagan, G., and S. P. Neuman (Eds.) (1997), Subsurface Flow and Transport: A Stochastic Approach, Cambridge Univ. Press, New York.

23. de Hoog, F. R., J. H. Knight, and A. N. Stokes (1982), An improved method for numerical inversion of Laplace transforms, SIAM J. Sci. Stat. Comput., 3, 357–366.

24. Deng, J., Jiang, X., Zhang, X., Hu, W., Crawford, J.W., 2008. Continuous time random walk model better describes the tailing of atrazine transport in soil. Chemosphere 71, 2150–2157.

25. Dentz, M., A. Cortis, H. Scher, and B. Berkowitz (2004), Time behavior of solute transport in heterogeneous media: Transition from anomalous to normal transport, Adv. Water Resour., 27, 155–173.

26. Dentz, M., H. Kinzelbach, S. Attinger, and W. Kinzelbach. 2002. Temporal behavior of a solute cloud in a heterogeneous porous medium—3. Numerical simulations. Water Resour. Res. 38(7):10.1029.

27. Devis, D., 1977. Statistic and analysis of geological data. Mir, 1977, page 572

28. Domenico, P.A. and F.W. Schwartz, 1990. Physical and Chemical Hydrogeology, John Wiley & Sons, New York, 824 p.

29. Domenico, P. A., and Robbins, G. A., 1984. A dispersion scale effect in model calibrations and field tracer experiments. J. Hydrol., 70:123-132.

30. Edery, Y., Scher, H., Berkowitz, B., 2009. Modeling bimolecular reactions and transport in porous media. Geophysical Research Letters 36, L02407.

31. Edery, Y., Scher, H., Berkowitz, B., 2010. Particle tracking model of bimolecular reactive transport in porous media. Water Resources Research 46.

32. Eggleston, J., and S. Rojstaczer (1998), Identification of large-scale hydraulic conductivity trends and the influence of trends on contaminant transport, Water Resour. Res., 34(9), 2155–2168.

33. Fried, J. J., 1975. Ground water pollution. Elsevier Scientific Publishing Company, Amsterdam – Oxford, New york.

34. Gaber, H.M., Inskeep, W.P., Comfort, S.D., Wraith, J.M., 1995. Nonequilibrium transport of atrazine through large intact soil cores. Soil Science Society of America Journal 59, 60–67.

35. Ghodrati, M., and W. A. Jury (1992), A field study of the effects of water application method and surface preparation method on preferential flow of pesticides in unsaturated soil, J. Contam. Hydrol., 11, 101–125.

36. G. Margolin, B. Berkowitz, J. Phys. Chem. B 104 (16) (2000) 3942, Minor correction: J. Phys. Chem. B, 104(36) 8762.

37. Hatano, Y., Hatano, N., 1998. Dispersive transport of ions in column experiments: an explanation of long-tailed profiles. Water Resources Research 34 (5), 1027–1033.

38. Hemker, 1997. Finite-Element Computer Program for Multiple-AquiferSteady-State and Transient Groundwater Flow Modelling. Amsterdam,The Netherlands.

39. Hoffman, F., D. Ronen, and Z. Pearl (1996), Evaluation of flow characteristics of a sand column using magnetic resonance imaging, J. Contam. Hydrol., 22:95–107.

40. Hughes, J. D., and Liu, J.: MIKE SHE: Software for Integrated Surface Water/Ground Water Modeling, Ground water, 46, 797-802, 2008.

41. Huang, K., Toride, N., van Genuchten, M. Th., 1995. Experimental investigation of solute transport in large, homogeneous and heterogeneous, saturated soil columns. Transport in porous media. 18: 283-302.

42. Ivakhnenko, A.G., 1975. Longtime prediction and proceedings of complex systems. Technika, page. 312 .

43. Jardine, P.M., G.K. Jacobs, and G.V. Wilson. 1993. Unsaturated transport processes in undisturbed heterogeneous porous media: I. Inorganic contaminants. Soil Sci. Soc. Am. J. 57:945–953.

44. Kenkre, V. M., E. W. Montroll, and M. F. Shlesinger (1973), Generalized master equations for continuous-time random walks, J. Stat. Phys., 9(1), 45–50.

45. Khan, A. U. H., and Jury, W. A., 1990, A laboratory study of the dispersion scale effect in column outflow experiments, J. Contaminant Hydrology 5, 119-131.

46. Kosakowski, G., B. Berkowitz, and H. Scher. 2001. Analysis of field observations of tracer transport in a fractured till. J. Contam. Hy-. drol. 47:29–51.

47. Le Borgne, T., Dentz,M., Carrera, J., 2008. Lagrangian statistical model for transport in highly heterogeneous velocity fields. Physical ReviewLetters 101.

48. Levy, M., and B. Berkowitz. 2003. Measurement and analysis of non- Fickian dispersion in heterogeneous porous media. J. Contam. - Hydrol. 64:203–226.

49. Li, N., Ren, L., 2009. Application of continuous time random walk theory to nonequilibrium transport in soil. Journal of Contaminant Hydrology 108, 134–151.

50. Lukner, L., 1986. Modelling of groundwater migration. Nedra, 1986, page 208

51. Lu, S.L., F.J. Molz, and G.J. Fix. 2002. Possible problems of scale dependency in applications of the three-dimensional fractional advection-dispersion equation to natural porous media. Water Resour. Res. 31: sour. Res. 38(9):1165.

52. Maraqa, M.A., Wallace, R.B., Voice, T.C., 1999. Effects of residence time and degree of water saturation on sorption nonequilibrium parameters. Journal of Contaminant Hydrology 36, 53–72.

53. Margolin, G., M. Dentz, and B. Berkowitz (2003), Continuous time random walk and multirate mass transfer modeling of sorption, Chem. Phys., 295, 71–80.

54. McLaughlin, D., and F. Ruan. 2001. Macrodispersivity and large-scale hydrogeologic variability. Transport Porous Med. 42:133–154.

55. Metzler, R., and J. Klafter. 2000. The random walk's guide to anomalous diffusion: A fractional dynamics approach. Phys. Rep. 339:1–77.

56. Metzler, R., and J. Klafter (2004), The restaurant at the end of the random walk: Recent developments in fractional dynamics of anomalous transport processes, J. Phys. A, 37, R161–R208.

57. Metzler, R. (2000), Generalized Chapman-Kolmogorov equation: A unifying approach to the description of anomalous transport in external fields, Phys. Rev. E, 62(5), 6233–6245.

58. Metzler, R., J. Klafter, and I. M. Sokolov (1998), Anomalous transport in external fields: Continuous time random walks and fractional diffusion equations extended, Phys. Rev. E, 58(2), 1621–1633.

59. Montroll, E. W., and G. H. Weiss (1965), Random walks on lattices. II, J. Math. Phys., 6(2), 167–181.

60. Montroll, E. W., and H. Scher (1973), Random walks on lattices. IV. Continuous time random walks and influence of absorbing boundaries, J. Stat. Phys., 9(2), 101–135.

61. Naff, R., D.F. Haley, and E.A. Sudicky. 1998. High-resolution Monte Carlo simulation of flow and conservative transport in heterogeneous porous media: 2. Transport results. Water Resour. Res. 34: 679–697.

62. Nielsen, D.R., and J.W. Biggar. 1962. Miscible displacement in soils: III. Theoretical considerations. Soil Sci. Soc. Am. Proc. 26:216–221.

63. Nirmala Peter, E., C., 2007, "Physical and numerical modelling of one –dimensional contaminant transport with scale dependent dispersion", PhD thesis.

64. Nkedi-Kizza, P., Brusseau, M.L., Suresh, P., Reo, C., Hornsby, A.G., 1989. Nonequilibrium sorption during displacement of hydrophobic organic chemicals and 45Ca through soil columns with aqueous and mixed solvents. Environmental Science and Technology 23, 814–820.

65. Ogata, A. and Banks, R. B., 1961. A solution of the differential equation of longitudinal dispersion in porous media. U. S. Geological Survey Professional Paper-411A:1-7.

66. Ognianik, N.S., Rudakov, V.K., 1985. Protection of groundwater in technogenetic development condition. Kiev:Vy˘sca ˘skola, page 221

67. Ognianik, N.S., Paramonova, N.K., 1977. Arguments for point mapping at aproximation of parameter with demand error. Geol. journal, 1-2, pages 159-164

68. Oppenheim, I., K. E. Shuler, and G. H. Weiss (1977), Stochastic Processes in Chemical Phsyics: The Master Equation, MIT Press, Cambridge, Mass.

69. Mori, H. (1965), Transport collective motion and Brownian motion,Prog. Theor. Phys., 33(3), 423–455.

70. Oswald, S., W. Kinzelbach, A. Greiner, and G. Brix (1997), Observation of flow and transport processes in artificial porous media via magnetic resonance imaging in three dimensions, Geoderma, 80, 417–429.

71. Pachepsky, Y., D. Benson, and R. Rawls. 2000. Simulating scale dependent solute transport in soils with the fractional advective dispersive equation. Soil Sci. Soc. Am. J. 64:1234–1243.

72. Pachepsky, Y., David Benson, and Walter Rawls, 2000. Simulating scale-dependent solute transport in soils with the Fractional advective-dispersive equation. Soil Sci. Soc. Am. J. 64: 1234-1243.

73. Pang, L., and Hunt, B., 2001, Solutions and verification of a scale-dependent dispersion model. Journal of Contaminant Hydrology, 53(1-2), 21-39.

74. Pannone, M., and P.K. Kitanidis. 2001. Large-time spatial covariance of concentration of conservative solute and application to the cape cod tracer test. Transport Porous Med. 42:109–132.

75. Peaceman D.W., and H.H. Rachford, Jr., 1962. Numerical Calculation of multidimensional miscible Displacement, J. Soc. Petro. Engrs., 2(4), 327-339.

76. Porta, G.M., Riva, M., Guadagnini, A., 2012. Upscaling solute transport in porous media in the presence of an irreversible bimolecular reaction. Advances Water Resources 35, 151–162.

77. Rhodes, M.E., Bijeljic, B., Blunt,M.J., 2008. Pore-to-field simulation of single-phase transport using continuous time random walks. Advances in Water Resources 31, 1527–1539.

78. Rumer, R. R., 1962. Longitudinal dispersion in steady and unsteady flow. Journal of the Hydraulic Division, ASCE, HY4, 147-172.

79. Salandin, P., and V. Fiorotto. 1998. Solute transport in highly heterogeneous aquifers. Water Resour. Res. 34:949–961.

80. Saltelli, A., K. Chan and M. Scott, Eds., 2000, Handbook of Sensitivity Analysis, John Wiley & Sons publishers, Probability and Statistics series.

81. Scheidegger, A.E. 1959. An evaluation of the accuracy of the diffusivity equation for describing miscible displacement in porous media.p. 101–116. In Proc. Theory of Fluid Flow in Porous Media Conf., Univ. of Oklahoma.

82. Scher, H., and M. Lax (1973b), Stochastic transport in a disordered solid. II. Impurity conduction, Phys. Rev. B, 7, 4502–4519.

83. Scher, H., G. Margolin, and B. Berkowitz (2002a), Towards a unified framework for anomalous transport in heterogeneous media, Chem. Phys., 284, 349–359.

84. Scher, H., G. Margolin, R. Metzler, J. Klafter, and B. Berkowitz (2002b), The dynamical foundation of fractal stream chemistry: The origin of extremely long retention times, Geophys. Res. Lett., 29(5), 1061.

85. Schumer, R., D.A. Benson, and M.M. Meerschaert. 2003. Fractal mobile/immobile solute transport. Water Resour. Res. 39(10):1296.

86. Scher, H., G. Margolin, and B. Berkowitz. 2002. Towards a unified framework for anomalous transport in heterogeneous media. Chem. Phys. 284:349–359

87. Scher, H., and E. W. Montroll (1975), Anomalous transit time dispersion in amorphous solids, Phys. Rev. B, 12, 2455–2477.

88. Scher, H., and M. Lax (1973a), Stochastic transport in a disordered solid. I. Theory, Phys. Rev. B, 7, 4491–4502.

89. Scher, H., and M. Lax. 1973b. Stochastic transport in a disordered solid. II. Impurity conduction. Phys. Rev. B 7:4502–4519.

90. Shlesinger, M. F. (1988), Fractal time in condensed matter, Annu. Rev. Phys. Chem., 39, 269–290.

91. Shlesinger, M. F. (1974), Asymptotic solutions of continuous-time random walks, J. Stat. Phys.,), 421–434.

92. Shlesinger, M. F. (1988), Fractal time in condensed matter, Annu. Rev. Phys. Chem., 39, 269– 290.

93. Shackelford, C. D., 1989.Diffusion of contaminants through waste containment barriers.

Transportation Research Record No. 1219, Council, Washington, D. C., 169-182.

94. Shackelford, C. D., 1994. Critical concepts for column testing. J. Of Geotechnical Engineering, ASCE, 120(10): 1804-1828.

95. Shackelford, C. D., 1995. Analytical models for cumulative mass column testing. Geoenvironment 2000, ASCE Geotechnical Specially Publication No. 46, Y. B. Acar and D. E. Daniel, Eds., ASCE, New York, 355-372.

96. Shackelford, C. D., and Daniel, D. E., 1991. Diffusion in saturated soil: Background. J. of Geotechnical Engineering, ASCE, 117(3):467-484.

97. Shackelford, C. D., and Redmond, P. L., 1995. Solute break-through curves for processed kaolin at low flow rates. J. of Geotechnical Engineering, ASCE, 121(1):17-32.

98. Shackelford, C. D., and Rowe, R. K., 1998. Contaminant transport modelling. Proceedings of the Third international conference on Environmental Geotechnics, Balkema, Rotterdam, 939-956.

99. Shamir, U., and D.R.F. Harleman, 1967. Numerical Solutions for Dispersion in Porous Mediums, Water Resour. Res., 3, 557-581.

100. Silliman, S. E., and E. S. Simpson (1987), Laboratory evidence of the scale effect in dispersion of solutes in porous media, Water Resour. Res., 23(8), 1667–1673.

101. S. Panday and P. S. Huyakorn , "A State-of-the-Art Use of Vadose Zone Flow and Transport Equations and Numerical Techniques for Environmental Evaluations.". Vadose Zone Journal. Vol. 7, No. 2, pp. 610-631. May 2008.

102. Soutter, M., Musy, A., 1997. Pesticide leaching models: sensitivity analyzes and Monte-Carlo simulations using Latin hypercube sampling, Water Resour.

103. Strack, 1989. Multi-Layer Analytic Element Model.

104. Stone, H.L., and P.L.T. Brian, 1963. Numerical Solution of Convective Transport Problems, Am. Inst. Chem. Engin. Jour., 9, 681-683 .

105. Taylor, S. R., and Howard, K. W. F., 1987. A field study of scale-dependent dispersion in a sandy aquifer. Journal of Hydrology, 90, 11-17.

106. Tiedje, T. (1984), Information about band-tail states from time-off light experiments, in Semiconductors and Semimetals, Part C, vol. 21, pp. 207–238, Elsevier, New York.

107. Toride, N., Leij, F., and van Genuchten, M. Th., 1995. The CXTFIT code for estimating transport parameters from laboratory or field tracer experiments.Version 2,

Research Report 137, US Salinity Lab., Riverside, CA.

108. Trefry, M.G., Muffels, C., FEFLOW: A finite-element ground water flow and transport modeling tool, Ground Water 45(5) (2007), 525–528.

109. Twarakavi, N.K.C., H. Saito, J. Šimůnek, and M.Th . van Genuchten. 2008. A new approach to estimate soil hydraulic parameters using only soil water retention data. Soil Sci. Soc. Am. J. 72:471–479.

110. Valdes-Parada, F.J., Alvarez-Ramirez, J., 2010. On the effective diffusivity under chemical reaction in porous media. Chemical Engineering Science 65, 4100–4104.

111. Villermaux, J. (1974), Deformation of chromatographic peaks under the influence of mass transfer phenomena, J. Chromatogr. Sci., 12, 822–831.

112. Villermaux, J. (1987), Chemical engineering approach to dynamic modelling of linear chromatography, J. Chromatogr., 406, 11– 26.

113. Wang, J. C., Booker, J. R., Carter, J. P., 1998. Experimental investigation of contaminant transport in porous media. Research Report No. R776,Centre for Geotechnical Engineering, Department of Civil Engineering, The University of Sydney.

114. Willmann, M., Carrera, J., Sánchez-Vila, X., 2008. Transport up scaling in heterogeneous aquifers: what physical parameters control memory functions, Water Resources Research 44 .

115. Yate, S. R., 1990. An analytical solution for one-dimensional transport in heterogeneous media. Water Resources Res., 26, 2331-2338.

116. Zaltzberg, E.A., 1976. Statistical methods predictions of natural regime of the groundwater table. Nedra, page 99.

117. Zhang, H., Selim, H.M., 2006. Modelling the transport and retention of arsenic (V) in soils. Soil Science Society of America Journal 70, 1677– 1687.

118. Zhang, R.G., Qian, J.Z., Ma, L., Qin, H., 2009. Application of extension identification method in mine water-bursting source discrimination. Journal of China Coal Society, Vol. 34, pp. 33-38.

119. Zhang, X., Crawford, J.W., Young, I.M., 2008. Does pore water velocity affect the reaction rates of adsorptive solute transport in soils. Demonstration with pore-scale modelling. Journal of Contaminant Hydrology 57, 241–258.

120. Zhou, L., and H.M. Selim. 2003. Application of the fractional advection-dispersion

equation in porous media. Soil Sci. Soc. Am. J. 67:1079–1084.

121. Zwanzig, R. (1960), Ensemble method in the theory of irreversibility, J.Chem. Phys., 33(5), 1338 –1341.

i want morebooks!

Buy your books fast and straightforward online - at one of world's
fastest growing online book stores! Environmentally sound due to
Print-on-Demand technologies.

Buy your books online at

www.get-morebooks.com

Kaufen Sie Ihre Bücher schnell und unkompliziert online – auf einer
der am schnellsten wachsenden Buchhandelsplattformen weltweit!
Dank Print-On-Demand umwelt- und ressourcenschonend produzi-
ert.

Bücher schneller online kaufen

www.morebooks.de

 VDM Verlagsservicegesellschaft mbH
Heinrich-Böcking-Str. 6-8   Telefon: +49 681 3720 174   info@vdm-vsg.de
D - 66121 Saarbrücken   Telefax: +49 681 3720 1749   www.vdm-vsg.de

Made in the USA
San Bernardino, CA
19 February 2020

64671533R00071